STUDENT SOLUTIONS MANUAL

FOR

FINITE MATHEMATICS

FOR THE
MANAGERIAL,
LIFE,
AND
SOCIAL SCIENCES

SEVENTH EDITION

FOR

FINITE MATHEMATICS

FOR THE
MANAGERIAL,
LIFE,
AND
SOCIAL SCIENCES

SEVENTH EDITION

S.T. Tan
Stonehill College

THOMSON
BROOKS/COLE

Australia • Canada • Mexico • Singapore • Spain • United Kingdom • United States

COPYRIGHT © 2003 Brooks/Cole, a division of Thomson Learning, Inc. Thomson Learning™ is a trademark used herein under license.

ALL RIGHTS RESERVED. No part of this work covered by the copyright hereon may be reproduced or used in any form or by any means—graphic, electronic, or mechanical, including but not limited to photocopying, recording, taping, Web distribution, information networks, or information storage and retrieval systems—without the written permission of the publisher.

Printed in Canada
1 2 3 4 5 6 7 06 05 04 03 02

Printer: Webcom Limited

ISBN: 0-534-39422-1

For more information about our products,
contact us at:
Thomson Learning Academic Resource Center
1-800-423-0563

For permission to use material from this text,
contact us by:
Phone: 1-800-730-2214
Fax: 1-800-731-2215
Web: http://www.thomsonrights.com

Asia
Thomson Learning
5 Shenton Way #01-01
UIC Building
Singapore 068808

Australia
Nelson Thomson Learning
102 Dodds Street
South Street
South Melbourne, Victoria 3205
Australia

Canada
Nelson Thomson Learning
1120 Birchmount Road
Toronto, Ontario M1K 5G4
Canada

Europe/Middle East/South Africa
Thomson Learning
High Holborn House
50/51 Bedford Row
London WC1R 4LR
United Kingdom

Latin America
Thomson Learning
Seneca, 53
Colonia Polanco
11560 Mexico D.F.
Mexico

Spain
Paraninfo Thomson Learning
Calle/Magallanes, 25
28015 Madrid, Spain

Cover image: © Cassandra Chu

FINITE MATHEMATICS
For The Managerial, Life, and Social Sciences

CONTENTS

CHAPTER 5 MATHEMATICS OF FINANCE

CHAPTER 6 SETS AND COUNTING

CHAPTER 7 PROBABILITY

CHAPTER 8 PROBABILITY DISTRIBUTIONS AND STATISTICS

CHAPTER 9 MARKOV CHAINS AND THE THEORY OF GAMES

CHAPTER 1

EXERCISES 1.1, page 8

1. The coordinates of A are (3,3) and it is located in Quadrant I.

3. The coordinates of C are (2,-2) and it is located in Quadrant IV.

5. The coordinates of E are (-4,-6) and it is located in Quadrant III.

7. A 9. $E, F,$ and G 11. F

For Exercises 13-19, refer to the following figure.

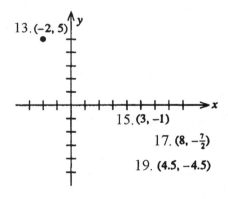

21. Using the distance formula, we find that $\sqrt{(4-1)^2 + (7-3)^2} = \sqrt{3^2 + 4^2} = \sqrt{25} = 5$.

23. Using the distance formula, we find that
$$\sqrt{(4-(-1))^2 + (9-3)^2} = \sqrt{5^2 + 6^2} = \sqrt{25+36} = \sqrt{61}.$$

25. The coordinates of the points have the form $(x,-6)$. Since the points are 10 units away from the origin, we have
$$(x-0)^2 + (-6-0)^2 = 10^2$$
$$x^2 = 64,$$
or $x = \pm 8$. Therefore, the required points are $(-8,-6)$ and $(8,-6)$.

27. The points are shown in the diagram that follows.

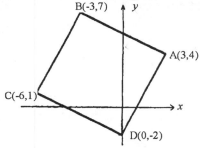

To show that the four sides are equal, we compute the following:

$$d(A,B) = \sqrt{(-3-3)^2 + (7-4)^2} = \sqrt{(-6)^2 + 3^2} = \sqrt{45}$$

$$d(B,C) = \sqrt{[(-6-(-3)]^2 + (1-7)^2} = \sqrt{(-3)^2 + (-6)^2} = \sqrt{45}$$

$$d(C,D) = \sqrt{[0-(-6)]^2 + [(-2)-1]^2} = \sqrt{(6)^2 + (-3)^2} = \sqrt{45}$$

$$d(A,D) = \sqrt{(0-3)^2 + (-2-4)^2} = \sqrt{(3)^2 + (-6)^2} = \sqrt{45}.$$

Next, to show that $\triangle ABC$ is a right triangle, we show that it satisfies the Pythagorean Theorem. Thus,

$$d(A,C) = \sqrt{(-6-3)^2 + (1-4)^2} = \sqrt{(-9)^2 + (-3)^2} = \sqrt{90} = 3\sqrt{10}$$

and $[d(A,B)]^2 + [d(B,C)]^2 = 90 = [d(A,C)]^2$. Similarly, $d(B,D) = \sqrt{90} = 3\sqrt{10}$, so $\triangle BAD$ is a right triangle as well. It follows that $\angle B$ and $\angle D$ are right angles, and we conclude that $ADCB$ is a square

29. The equation of the circle with radius 5 and center (2,-3) is given by

$$(x-2)^2 + [y-(-3)]^2 = 5^2$$

or $$(x-2)^2 + (y+3)^2 = 25.$$

31. The equation of the circle with radius 5 and center (0, 0) is given by

$$(x-0)^2 + (y-0)^2 = 5^2 \qquad \text{or} \qquad x^2 + y^2 = 25$$

33. The distance between the points (5,2) and (2,-3) is given by

$$d = \sqrt{(5-2)^2 + (2-(-3))^2} = \sqrt{3^2 + 5^2} = \sqrt{34}.$$

Therefore $r = \sqrt{34}$ and the equation of the circle passing through (5,2) and centered at (2,-3) is $(x-2)^2 + [y-(-3)]^2 = 34$, or $\qquad (x-2)^2 + (y+3)^2 = 34$.

35. Referring to the diagram on page 9 of the text, we see that the distance from A to B is given by $d(A,B) = \sqrt{400^2 + 300^2} = \sqrt{250,000} = 500$. The distance from B to C is given by

$$d(B,C) = \sqrt{(-800-400)^2 + (800-300)^2} = \sqrt{(-1200)^2 + (500)^2}$$
$$= \sqrt{1,690,000} = 1300.$$

The distance from C to D is given by

$$d(C,D) = \sqrt{[-800-(-800)]^2 + (800-0)^2} = \sqrt{0+800^2} = 800 .$$

The distance from D to A is given by

$$d(D,A) = \sqrt{[0-(-800)]^2 + (0-0)} = \sqrt{640000} = 800.$$

Therefore, the total distance covered on the tour, is

$$d(A,B)+d(B,C)+d(C,D)+d(D,A) = 500+1300+800+800$$
$$= 3400, \quad \text{or} \ 3400 \ \text{miles.}$$

37. Referring to the following diagram,

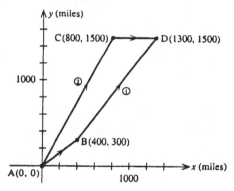

we see that the distance he would cover if he took Route (1) is given by

$$d(A,B)+d(B,D) = \sqrt{400^2+300^2} + \sqrt{(1300-400)^2 + (1500-300)^2}$$
$$= \sqrt{250,000} + \sqrt{2,250,000} = 500+1500 = 2000,$$

or 2000 miles. On the other hand, the distance he would cover if he took Route (2) is given by

$$d(A,C)+d(C,D) = \sqrt{800^2+1500^2} + \sqrt{(1300-800)^2}$$
$$= \sqrt{2,890,000} + \sqrt{250,000} = 1700+500 = 2200,$$

or 2200 miles. Comparing these results, we see that he should take Route (1).

39. Calculations to determine VHF requirements:
$$d = \sqrt{25^2 + 35^2} \ = \sqrt{625 + 1225} = \sqrt{1850} \approx 43.01.$$
Models B through D satisfy this requirement.
Calculations to determine UHF requirements:
$$d = \sqrt{20^2 + 32^2} = \sqrt{400 + 1024} = \sqrt{1424} \approx 37.74$$
Models C through D satisfy this requirement. Therefore, Model C will allow him to receive both channels at the least cost.

41. a. Let the position of ship A and ship B after t hours be $A(0, y)$ and $B(x, 0)$, respectively. Then $x = 30t$ and $y = 20t$. Therefore, the distance between the two ships is
$$D = \sqrt{(30t)^2 + (20t)^2} = \sqrt{900t^2 + 400t^2} = 10\sqrt{13}t.$$
b. The required distance is obtained by letting $t = 2$ giving $D = 10\sqrt{13}(2)$ or approximately 72.11 miles.

43. False. The distance between $P_1(a,b)$ and $P_3(kc,kd)$ is
$$d = \sqrt{(kc - a)^2 + (kd - b)^2}$$
$$\neq |k| D = |k| \sqrt{(c - a)^2 + (d - b)^2} = \sqrt{k^2(c - a)^2 + k^2(d - b)^2} = \sqrt{[k(c - a)]^2 + [k(d - b)]^2}$$

45. Referring to the figure in the text, we see that the distance between the two points is given by the length of the hypotenuse of the right triangle. That is,
$$d = \sqrt{(x_2 - x_1)^2 + (y_2 - y_1)^2}$$

EXERCISES 1.2, page 21

1. Referring to the figure shown in the text, we see that $m = \dfrac{2 - 0}{0 - (-4)} = \dfrac{1}{2}$.

3. This is a vertical line, and hence its slope is undefined.

5. $m = \dfrac{y_2 - y_1}{x_2 - x_1} = \dfrac{8 - 3}{5 - 4} = 5.$ 7. $m = \dfrac{y_2 - y_1}{x_2 - x_1} = \dfrac{8 - 3}{4 - (-2)} = \dfrac{5}{6}.$

9. $m = \dfrac{y_2 - y_1}{x_2 - x_1} = \dfrac{d - b}{c - a}.$

11. Since the equation is in the slope-intercept form, we read off the slope $m = 4$.

   a. If x increases by 1 unit, then y increases by 4 units.

   b. If x decreases by 2 units, y decreases by $4(-2) = -8$ units.

13. The slope of the line through A and B is $\dfrac{-10-(-2)}{-3-1} = \dfrac{-8}{-4} = 2$.

   The slope of the line through C and D is $\dfrac{1-5}{-1-1} = \dfrac{-4}{-2} = 2$.

   Since the slopes of these two lines are equal, the lines are parallel.

15. The slope of the line through A and B is $\dfrac{2-5}{4-(-2)} = -\dfrac{3}{6} = -\dfrac{1}{2}$.

   The slope of the line through C and D is $\dfrac{6-(-2)}{3-(-1)} = \dfrac{8}{4} = 2$.

   Since the slopes of these two lines are the negative reciprocals of each other, the lines are perpendicular.

17. The slope of the line through the point $(1, a)$ and $(4,-2)$ is $m_1 = \dfrac{-2-a}{4-1}$ and the slope of the line through $(2,8)$ and $(-7, a + 4)$ is $m_2 = \dfrac{a+4-8}{-7-2}$. Since these two lines are parallel, m_1 is equal to m_2. Therefore,

$$\frac{-2-a}{3} = \frac{a-4}{-9}$$
$$-9(-2-a) = 3(a-4)$$
$$18+9a = 3a-12$$
$$6a = -30 \qquad \text{and} \quad a = -5$$

19. An equation of a horizontal line is of the form $y = b$. In this case $b = -3$, so $y = -3$ is an equation of the line.

21. e $\qquad\qquad$ 23. a $\qquad\qquad$ 25. f

27. We use the point-slope form of an equation of a line with the point $(3,-4)$ and slope $m = 2$. Thus $\qquad y - y_1 = m(x - x_1)$,

and
$$y - (-4) = 2(x - 3)$$
$$y + 4 = 2x - 6$$
$$y = 2x - 10.$$

29. Since the slope $m = 0$, we know that the line is a horizontal line of the form $y = b$. Since the line passes through $(-3,2)$, we see that $b = 2$, and an equation of the line is $y = 2$.

31. We first compute the slope of the line joining the points $(2,4)$ and $(3,7)$. Thus,
$$m = \frac{7-4}{3-2} = 3.$$
Using the point-slope form of an equation of a line with the point $(2,4)$ and slope $m = 3$, we find
$$y - 4 = 3(x - 2)$$
$$y = 3x - 2.$$

33. We first compute the slope of the line joining the points $(1,2)$ and $(-3,-2)$. Thus,
$$m = \frac{-2-2}{-3-1} = \frac{-4}{-4} = 1.$$
Using the point-slope form of an equation of a line with the point $(1,2)$ and slope $m = 1$, we find
$$y - 2 = x - 1$$
$$y = x + 1.$$

35. We use the slope-intercept form of an equation of a line: $y = mx + b$. Since $m = 3$, and $b = 4$, the equation is $y = 3x + 4$.

37. We use the slope-intercept form of an equation of a line: $y = mx + b$. Since $m = 0$, and $b = 5$, the equation is $y = 5$.

39. We first write the given equation in the slope-intercept form:
$$x - 2y = 0$$
$$-2y = -x$$
$$y = \tfrac{1}{2}x \ .$$
From this equation, we see that $m = 1/2$ and $b = 0$.

41. We write the equation in slope-intercept form:

$$2x - 3y - 9 = 0$$
$$-3y = -2x + 9$$
$$y = \tfrac{2}{3}x - 3.$$

From this equation, we see that $m = 2/3$ and $b = -3$.

43. We write the equation in slope-intercept form:
$$2x + 4y = 14$$
$$4y = -2x + 14$$
$$y = -\tfrac{2}{4}x + \tfrac{14}{4}$$
$$= -\tfrac{1}{2}x + \tfrac{7}{2}.$$

From this equation, we see that $m = -1/2$ and $b = 7/2$.

45. We first write the equation $2x - 4y - 8 = 0$ in slope- intercept form:
$$2x - 4y - 8 = 0$$
$$4y = 2x - 8$$
$$y = \tfrac{1}{2}x - 2$$

Now the required line is parallel to this line, and hence has the same slope. Using the point-slope equation of a line with $m = 1/2$ and the point $(-2,2)$, we have
$$y - 2 = \tfrac{1}{2}[x - (-2)] = \tfrac{1}{2}x + 1$$
$$y = \tfrac{1}{2}x + 3.$$

47. A line parallel to the x-axis has slope 0 and is of the form $y = b$. Since the line is 6 units below the axis, it passes through $(0,-6)$ and its equation is $y = -6$.

49. We use the point-slope form of an equation of a line to obtain
$$y - b = 0(x - a) \quad \text{or} \quad y = b.$$

51. Since the required line is parallel to the line joining $(-3,2)$ and $(6,8)$, it has slope
$$m = \frac{8 - 2}{6 - (-3)} = \frac{6}{9} = \frac{2}{3}.$$

We also know that the required line passes through $(-5,-4)$. Using the point-slope form of an equation of a line, we find
$$y - (-4) = \tfrac{2}{3}(x - (-5)); \quad y = \tfrac{2}{3}x + \tfrac{10}{3} - 4, \text{ or } y = \tfrac{2}{3}x - \tfrac{2}{3}$$

53. Since the point $(-3,5)$ lies on the line $kx + 3y + 9 = 0$, it satisfies the equation.

Substituting $x = -3$ and $y = 5$ into this equation gives
$$-3k + 15 + 9 = 0 \quad \text{or} \quad k = 8.$$

55. $3x - 2y + 6 = 0$

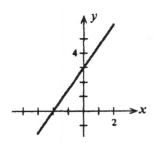

57. $x + 2y - 4 = 0$

59. $y + 5 = 0$

61. Since the line passes through the points $(a, 0)$ and $(0, b)$, its slope is $m = \dfrac{b - 0}{0 - a} = -\dfrac{b}{a}$.

Then, using the point-slope form of an equation of a line with the point $(a, 0)$ we have
$$y - 0 = -\tfrac{b}{a}(x - a)$$
$$y = -\tfrac{b}{a}x + b$$

which may be written in the form $\tfrac{b}{a}x + y = b$.

Multiplying this last equation by $1/b$, we have $\quad \dfrac{x}{a} + \dfrac{y}{b} = 1$.

63. Using the equation $\dfrac{x}{a} + \dfrac{y}{b} = 1$ with $a = -2$ and $b = -4$, we have $-\dfrac{x}{2} - \dfrac{y}{4} = 1$.

Then
$$-4x - 2y = 8$$
$$2y = -8 - 4x$$
$$y = -2x - 4.$$

65. Using the equation $\dfrac{x}{a} + \dfrac{y}{b} = 1$ with $a = 4$ and $b = -1/2$, we have

$$\frac{x}{4} + \frac{y}{-\frac{1}{2}} = 1$$

$$-\tfrac{1}{4}x + 2y = -1$$

$$2y = \tfrac{1}{4}x - 1, \quad y = \tfrac{1}{8}x - \tfrac{1}{2}.$$

67. The slope of the line passing through A and B is $m = \dfrac{7-1}{1-(-2)} = \dfrac{6}{3} = 2$,

and the slope of the line passing through B and C is $m = \dfrac{13-7}{4-1} = \dfrac{6}{3} = 2$.

Since the slopes are equal, the points lie on the same line.

69. a.

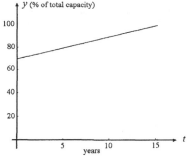

b. The slope is 1.9467 and the y-intercept is 70.082.
c. The output is increasing at the rate of 1.9467%/yr; the output at the beginning of 1990 was 70.082%.
d. We solve the equation $1.9467t + 70.082 = 100$ giving $t = 15.37$. We conclude that the plants will be generating at maximum capacity shortly after 2005.

71. a. $y = 0.55x$

b. Solving the equation $1100 = 0.55x$ for x, we have $x = \dfrac{1100}{0.55} = 2000$.

73. Using the points $(0, 0.68)$ and $(10, 0.80)$, we see that the slope of the required line is

$$m = \frac{0.80 - 0.68}{10 - 0} = \frac{0.12}{10} = .012.$$

Next, using the point-slope form of the equation of a line, we have
$$y - 0.68 = 0.012(t - 0)$$
or $\qquad y = 0.012t + 0.68.$

Therefore, when $t = 14$, we have $y = 0.012(14) + 0.68 = .848$

or 84.8%. That is, in 2004 women's wages are expected to be 84.8% of men's wages

75. a. – b.

c. Using the points (0,200) and (100,250), we see that the slope of the required line

is $m = \dfrac{250 - 200}{100} = \dfrac{1}{2}$. Therefore, the required equation is

$$y - 200 = \tfrac{1}{2}x \quad \text{or} \quad y = \tfrac{1}{2}x + 200.$$

d. The approximate cost for producing 54 units of the commodity is
$\tfrac{1}{2}(54) + 200, \quad \text{or } \$227.$

77. a. – b.

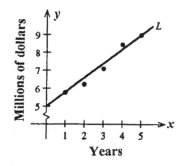

c. The slope of L is $m = \dfrac{9.0 - 5.8}{5 - 1} = \dfrac{3.2}{4} = 0.8$. Using the point-slope form of an

equation of a line, we have $y - 5.8 = 0.8(x - 1) = 0.8x - 0.8$, or $\quad y = 0.8x + 5$.

d. Using the equation of part (c) with $x = 9$, we have

$$y = 0.8(9) + 5 = 12.2, \quad \text{or } \$12.2 \text{ million.}$$

79. a. We obtain a family of parallel lines each having slope m.
    b. We obtain a family of straight lines all of which pass through the point $(0,b)$.

81. False. Substituting $x = -1$ and $y = 1$ into the equation gives $3(-1) + 7(1) = 4$, and this is not equal to the right-hand side of the equation. Therefore, the equation is not satisfied and so the given point does not lie on the line.

83. True. The slope of the line $Ax + By + C = 0$ is $-A/B$. (Write it in the slope-intercept form.) Similarly, the slope of the line $ax + by + c = 0$ is $-a/b$. They are parallel if and only if $-\dfrac{A}{B} = -\dfrac{a}{b}$, $Ab = aB$, or $Ab - aB = 0$.

85. True. The slope of the line $ax + by + c_1 = 0$ is $m_1 = -a/b$. The slope of the line $bx - ay + c_2 = 0$ is $m_2 = b/a$. Since $m_1 m_2 = -1$, the straight lines are indeed perpendicular.

87. Writing each equation in the slope-intercept form, we have
$$y = -\frac{a_1}{b_1}x - \frac{c_1}{b_1} \quad (b_1 \neq 0) \quad \text{and} \quad y = -\frac{a_2}{b_2}x - \frac{c_2}{b_2} \quad (b_2 \neq 0)$$
Since two lines are parallel if and only if their slopes are equal, we see that the lines are parallel if and only if $-\dfrac{a_1}{b_1} = -\dfrac{a_2}{b_2}$, or $a_1b_2 - b_1a_2 = 0$.

USING TECHNOLOGY EXERCISES 1.2, page 27

1.

3.

5.

7. a.

b.

9. a.

b.

11.

13.

15.   17.

EXCEL

1.

3.

5.

7.

9.

11.

13.

15.

17.

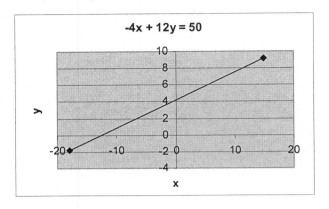

-4x + 12y = 50

EXERCISES 1.3, page 37

1. Yes. Solving for y in terms of x, we find $3y = -2x + 6$, or $y = -\frac{2}{3}x + 2$.

3. Yes. Solving for y in terms of x, we find $2y = x + 4$, or $y = \frac{1}{2}x + 2$.

5. Yes. Solving for y in terms of x, we have $4y = 2x + 9$, or $y = \frac{1}{2}x + \frac{9}{4}$.

7. y is not a linear function of x because of the quadratic term $2x^2$.

9. y is not a linear function of x because of the nonlinear term $-3y^2$.

11. a. $C(x) = 8x + 40,000$, where x is the number of units produced.
    b. $R(x) = 12x$, where x is the number of units sold.
    c. $P(x) = R(x) - C(x) = 12x - (8x + 40,000) = 4x - 40,000$.
    d. $P(8,000) = 4(8,000) - 40,000 = -8,000$, or a loss of \$8,000.
       $P(12,000) = 4(12,000) - 40,000 = 8,000$ or a profit of \$8,000.

13. $f(0) = 2$ gives $m(0) + b = 2$, or $b = 2$. So, $f(x) = mx + 2$. Next, $f(3) = -1$ gives $m(3) + 2 = -1$, or $m = -1$.

15. Let V be the book value of the office building after 2000. Since $V = 1,000,000$ when $t = 0$, the line passes through $(0, 1,000,000)$. Similarly, when $t = 50$, $V = 0$, so the line passes through $(50, 0)$. Then the slope of the line is given by

$$m = \frac{0 - 1,000,000}{50 - 0} = -20,000.$$

Using the point-slope form of the equation of a line with the point (0, 1,000,000), we have $\quad V - 1,000,000 = -20,000(t - 0),\quad$ or $\quad V = -20,000t + 1,000,000.$

In 2005, $t = 5$ and $\quad V = -20,000(5) + 1,000,000 = 900,000,\quad$ or \$900,000.

In 2010, $t = 10$ and $V = -20,000(10) + 1,000,000 = 800,000,\quad$ or \$800,000.

17. The consumption function is given by $C(x) = 0.75x + 6$. When $x = 0$, we have $C(0) = 0.75(0) + 6 = 6$, or \$6 billion dollars.

When $x = 50$, $\qquad C(50) = 0.75(50) + 6 = 43.5,\quad$ or \$43.5 billion dollars.

When $x = 100$, $\quad C(100) = 0.75(100) + 6 = 81,\quad$ or \$81 billion dollars.

19. a. $y = 1.053x$, where x is the monthly benefit before adjustment, and y is the adjusted monthly benefit.

b. His adjusted monthly benefit will be $(1.053)(620) = 652.86$, or \$652.86.

21. Let the number of tapes produced and sold be x. Then
$$C(x) = 12,100 + 0.60x; \qquad R(x) = 1.15x$$
and $\qquad P(x) = R(x) - C(x) = 1.15x - (12,100 + 0.60x) = 0.55x - 12,100.$

23. Let the value of the server after t years be V. When $t = 0$, $V = 60,000$ and when $t = 4$, $V = 12,000$.

a. Since $\qquad m = \dfrac{12,000 - 60,000}{4} = -\dfrac{48,000}{4} = -12,000$

the rate of depreciation $(-m)$ is \$12,000.

b. Using the point-slope form of the equation of a line with the point (4, 12,000), we have $\qquad V - 12,000 = -12,000(t - 4)$

or $\qquad\qquad\qquad V = -12,000t + 60,000.$

c.

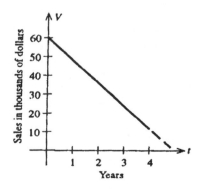

17   *1 Straight Lines and Linear Functions*

d. When $t = 3$, $V = -12,000(3) + 60,000 = 24,000$, or \$24,000.

25. The formula given in Exercise 24 is $\qquad V = C - \dfrac{C-S}{N}t$.

    Then, when $C = 1,000,000$, $N = 50$, and $S = 0$, we have
    $$V = 1,000,000 - \frac{1,000,000 - 0}{50}t \quad \text{or} \quad V = 1,000,000 - 20,000t.$$
    In 2005, $t = 5$ and $\quad V = 1,000,000 - 20,000(5) = 900,000$, or \$900,000.
    In 2010, $t = 10$ and $\quad V = 1,000,000 - 20,000(10) = 800,000$ or \$800,000.

27. a. $D(S) = \dfrac{Sa}{1.7}$. If we think of D as having the form $D(S) = mS + b$, then

    $m = \dfrac{a}{1.7}$, $b = 0$, and D is a linear function of S.

    b. $D(0.4) = \dfrac{500(0.4)}{1.7} = 117.647$, or approximately 117.65 mg.

29. a. Since the relationship is linear, we can write $F = mC + b$, where m and b are
    constants. Using the condition $C = 0$ when $F = 32$, we have $32 = b$, and so
    $F = mC + 32$. Next, using the condition $C = 100$ when $F = 212$, we have
    $212 = 100m + 32$ or $m = \frac{9}{5}$. Therefore, $F = \frac{9}{5}C + 32$.

    b. From (a), we see $F = \frac{9}{5}C + 32$. Next, when $C = 20$, $F = \frac{9}{5}(20) + 32 = 68$
    and so the temperature equivalent to 20°C is 68°F.

    c. Solving for C in terms of F, we find $\frac{9}{5}C = F - 32$, or $C = \frac{5}{9}F - \frac{160}{9}$.
    When $F = 70$, $C = \frac{5}{9}(70) - \frac{160}{9} = \frac{190}{9}$, or approximately 21.1°C.

31. The slope of L_2 is greater than that of L_1. This tells us that if the manufacturer lowers
    the unit price for each model clock radio by the same amount, the additional quantity
    demanded of model B radios will be greater than that of the model A radios.

33. a. Setting $x = 0$, gives $3p = 18$, or $p = 6$. Next, setting $p = 0$, gives $2x = 18$, or $x = 9$.

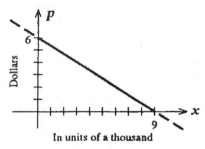

In units of a thousand

b. When $p = 4$, $\qquad 2x + 3(4) - 18 = 0$

$$2x = 18 - 12 = 6$$

and $x = 3$. Therefore, the quantity demanded when $p = 4$ is 3000. (Remember x is given in units of a thousand.)

35. a. When $x = 0$, $p = 60$ and when $p = 0$, $-3x = -60$, or $x = 20$.

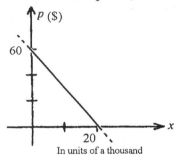

In units of a thousand

b. When $p = 30$, $\qquad 30 = -3x + 60$

$$3x = 30 \qquad\qquad \text{and} \quad x = 10.$$

Therefore, the quantity demanded when $p = 30$ is 10,000 units.

37. When $x = 1000$, $p = 55$, and when $x = 600$, $p = 85$. Therefore, the graph of the linear demand equation is the straight line passing through the points (1000, 55) and (600, 85). The slope of the line is

$$\frac{85 - 55}{600 - 1000} = -\frac{3}{40}.$$

Using the point (1000, 55) and the slope just found, we find that the required equation is $\quad p - 55 = -\frac{3}{40}(x - 1000)$

$$p = -\frac{3}{40}x + 130$$

When $x = 0$, $p = 130$ which means that there will be no demand above \$130.
When $p = 0$, $x = 1733.33$, which means that 1733 units is the maximum quantity demanded.

39. Since the demand equation is linear, we know that the line passes through the points (1000,9) and (6000,4). Therefore, the slope of the line is given by
$$m = \frac{4-9}{6000-1000} = -\frac{5}{5000} = -0.001 .$$
Since the equation of the line has the form $p = ax + b$,
$$9 = -0.001(1000) + b, \quad \text{or} \quad b = 10.$$
Therefore, the equation of the line is $p = -0.001x + 10$.
If $p = 7.50$, $\quad 7.50 = -0.001x + 10$
$$0.001x = 2.50, \quad \text{or} \quad x = 2500.$$
So, the quantity demanded when the unit price is \$7.50 is 2500 units.

41. a. Setting $x = 0$, we obtain $3(0) - 4p + 24 = 0$, $-4p = -24$, or $p = 6$.
Setting $p = 0$, we obtain $\quad 3x - 4(0) + 24 = 0$, $\quad 3x = -24$
or $\hspace{8em} x = -8.$

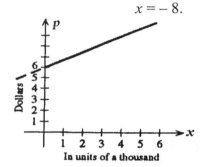

In units of a thousand

b. When $p = 8$, $\hspace{6em} 3x - 4(8) + 24 = 0$
$$3x = 32 - 24 = 8$$
$$x = 8/3.$$

Therefore, 2667 units of the commodity would be supplied at a unit price of \$8. (Here again x is measured in units of thousands.)

43. a. When $x = 0$, $p = 10$, and when $p = 0$, $x = -5$.

In units of a thousand

b. $p = 2x + 10$, $14 = 2x + 10$, $2x = 4$, and $x = 2$. Therefore, when $p = 14$, the supplier will make 2000 units of the commodity available.

45. When $x = 10{,}000$, $p = 45$ and when $x = 20{,}000$, $p = 50$. Therefore, the slope of the line passing $(10{,}000, 45)$ and $(20{,}000, 50)$ is

$$m = \frac{50 - 45}{20{,}000 - 10{,}000} = \frac{5}{10{,}000} = 0.0005$$

Using the point- slope form of an equation of a line with the point $(10{,}000, 45)$, we have $\qquad p - 45 = 0.0005(x - 10{,}000)$

$$p = 0.0005x - 5 + 45$$

or $\qquad\qquad p = 0.0005x + 40.$

If $p = 70$, $\qquad 70 = 0.0005x + 40$

$$0.0005x = 30, \qquad \text{or} \quad x = \frac{30}{0.0005} = 60{,}000$$

(If x is expressed in units of a thousand, then then the equation may be written in the form $p = \frac{1}{2}x + 40$.)

In units of a thousand

47. False. $P(x) = R(x) - C(x) = sx - (cx + F) = (s - c)x - F$. Therefore, the firm is making a profit if $P(x) = (s - c)x - F > 0$ or $x > \dfrac{F}{s - c}$.

1. 2.2875   3. 2.880952381   5. 7.2851648352   7. 2.4680851064

EXERCISES 1.4, page 51

1. We solve the system $y = \ \ 3x + 4$
    $$y = -2x + 14.$$
    Substituting the first equation into the second yields
    $$3x + 4 = -2x + 14$$
    $$5x = 10,$$
    and $x = 2$. Substituting this value of x into the first equation yields
    $$y = 3(2) + 4,$$
    or $y = 10$. Thus, the point of intersection is $(2,10)$.

3. We solve the system $2x - 3y = \ 6$
    $$3x + 6y = 16.$$
    Solving the first equation for y, we obtain
    $$3y = 2x - \ 6$$
    $$y = \tfrac{2}{3}x - 2 \ .$$
    Substituting this value of y into the second equation, we obtain
    $$3x + 6(\tfrac{2}{3}x - 2) = 16$$
    $$3x + 4x - \ 12 = 16$$
    $$7x = 28$$
    and $x = 4.$
    Then $y = \tfrac{2}{3}(4) - 2 = \tfrac{2}{3}.$
    Therefore, the point of intersection is $(4, \tfrac{2}{3})$.

5. We solve the system $\begin{cases} y = \tfrac{1}{4}x - 5 \\ 2x - \tfrac{3}{2}y = 1 \end{cases}$. Substituting the value of y given in the first

    equation into the second equation, we obtain
    $$2x - \tfrac{3}{2}(\tfrac{1}{4}x - 5) = 1$$
    $$2x - \tfrac{3}{8}x + \tfrac{15}{2} = 1$$
    $$16x - \ 3x + 60 = 8$$
    $$13x = -52,$$
    or $x = -4$. Substituting this value of x in the first equation, we have

1 Straight Lines and Linear Functions   **22**

$$y = \tfrac{1}{4}(-4) - 5 \; = -1 - \; 5,$$

or $y = -6$. Therefore, the point of intersection is $(-4, -6)$.

7. We solve the equation $R(x) = C(x)$, or $15x = 5x + 10{,}000$, obtaining $10x = 10{,}000$, or $x = 1000$. Substituting this value of x into the equation $R(x) = 15x$, we find $R(1000) = 15{,}000$. Therefore, the breakeven point is $(1000, 15{,}000)$.

9. We solve the equation $R(x) = C(x)$, or $0.4x = 0.2x + 120$, obtaining $0.2x = 120$, or $x = 600$. Substituting this value of x into the equation $R(x) = 0.4x$, we find $R(600) = 240$. Therefore, the breakeven point is $(600, 240)$.

11. a.

b. We solve the equation $R(x) = C(x)$ or $14x = 8x + 48{,}000$, obtaining $6x = 48{,}000$ or $x = 8000$. Substituting this value of x into the equation $R(x) = 14x$, we find

$R(8000) = 14(8000) = 112{,}000$. Therefore, the breakeven point is $(8000, 112{,}000)$.

c.

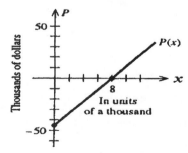

d. $P(x) = R(x) - C(x) = 14x - 8x - 48{,}000 = 6x - 48{,}000$.

The graph of the profit function crosses the x-axis when $P(x) = 0$, or $6x = 48,000$ and $x = 8000$. This means that the revenue is equal to the cost when 8000 units are produced and consequently the company breaks even at this point.

13. Let x denote the number of units sold. Then, the revenue function R is given by
$$R(x) = 9x.$$
Since the variable cost is 40 percent of the selling price and the monthly fixed costs are \$50,000, the cost function C is given by
$$C(x) = 0.4(9x) + 50,000$$
$$= 3.6x + 50,000.$$
To find the breakeven point, we set $R(x) = C(x)$, obtaining
$$9x = 3.6x + 50,000$$
$$5.4x = 50,000$$
$$x \approx 9259, \text{ or } 9259 \text{ units.}$$
Substituting this value of x into the equation $R(x) = 9x$ gives
$$R(9259) = 9(9259) = 83,331.$$
Thus, for a breakeven operation, the firm should manufacture 9259 bicycle pumps resulting in a breakeven revenue of \$83,331.

15. a. The cost function associated with using machine I is given by
$$C_1(x) = 18,000 + 15x.$$
The cost function associated with using machine II is given by
$$C_2(x) = 15,000 + 20x.$$

b.

In units of a hundred

c. Comparing the cost of producing 450 units on each machine, we find
$$C_1(450) = 18,000 + 15(450)$$
$$= 24,750 \quad \text{or } \$24,750 \text{ on machine } I,$$
and $\qquad C_2(450) = 15,000 + 20(450)$
$$= 24,000 \quad \text{or } \$24,000 \text{ on machine } II.$$

Therefore, machine *II* should be used in this case.

Next, comparing the costs of producing 550 units on each machine, we find

$$C_1(550) = 18,000 + 15(550)$$
$$= 26,250 \text{ or } \$26,250 \text{ on machine } I,$$

and $\quad C_2(550) = 15,000 + 20(550)$
$$= 26,000$$

on machine *II*. Therefore, machine *II* should be used in this instance. Once again, we compare the cost of producing 650 units on each machine and find that

$$C_1(650) = 18,000 + 15(650)$$
$$= 27,750, \text{ or } \$27,750 \text{ on machine } I \text{ and}$$

$$C_2(650) = 15,000 + 20(650)$$
$$= 28,000,$$

or $28,000 on machine *II*. Therefore, machine *I* should be used in this case.

d. We use the equation $P(x) = R(x) - C(x)$ and find

$$P(450) = 50(450) - 24,000 = -1500,$$

or a loss of $1500 when machine *I* is used to produce 450 units. Similarly,

$$P(550) = 50(550) - 26,000 = 1500,$$

or a profit of $1500 when machine *II* is used to produce 550 units.

Finally, $P(650) = 50(650) - 27,750 = 4750,$

or a profit of $4750 when machine *I* is used to produce 650 units.

17. We solve the system

$$4x + 3p = 59$$
$$5x - 6p = -14.$$

Solving the first equation for p, we find $p = -\frac{4}{3}x + \frac{59}{3}$.

Substituting this value of p into the second equation, we have

$$5x - 6(-\tfrac{4}{3}x + \tfrac{59}{3}) = -14$$
$$5x + 8x - 118 = -14$$
$$13x = 104$$
$$x = 8.$$

Substituting this value of x into the equation

$$p = -\tfrac{4}{3}x + \tfrac{59}{3}$$

we have $\qquad p = -\tfrac{4}{3}(8) + \tfrac{59}{3} = \tfrac{27}{3} = 9$

Thus, the equilibrium quantity is 8000 units and the equilibrium price is $9.

19. We solve the system $\quad p = -2x + 22$
$$p = 3x + 12$$

Substituting the first equation into the second, we find
$$-2x + 22 = 3x + 12$$
$$5x = 10$$
and
$$x = 2.$$
Substituting this value of x into the first equation, we obtain
$$p = -2(2) + 22 = 18.$$
Thus, the equilibrium quantity is 2000 units and the equilibrium price is \$18.

21. Let x denote the number of DVD's produced per week, and p denote the price of each DVD.

a. The slope of the demand curve is given by $\dfrac{\Delta p}{\Delta x} = -\dfrac{20}{250} = -\dfrac{2}{25}$.

Using the point-slope form of the equation of a line with the point (3000, 485), we have
$$p - 485 = -\tfrac{2}{25}(x - 3000)$$
$$p = -\tfrac{2}{25}x + 240 + 485$$
or
$$p = -0.08x + 725.$$
b. From the given information, we know that the graph of the supply equation passes through the points (0, 300) and (2500, 525). Therefore, the slope of the supply curve

is
$$m = \dfrac{525 - 300}{2500 - 0} = \dfrac{225}{2500} = 0.09.$$

Using the point-slope form of the equation of a line with the point (0, 300), we find that
$$p - 300 = 0.09x$$
$$p = 0.09x + 300.$$
c. Equating the supply and demand equations, we have
$$-0.08x + 725 = 0.09x + 300$$
$$0.17x = 425$$
or
$$x = 2500.$$
Then
$$p = -0.08(2500) + 725 = 525.$$
We conclude that the equilibrium quantity is 2500 and the equilibrium price is \$525.

23. We solve the system
$$3x + p = 1500$$
$$2x - 3p = -1200.$$
Solving the first equation for p, we obtain
$$p = 1500 - 3x.$$
Substituting this value of p into the second equation, we obtain
$$2x - 3(1500 - 3x) = -1200$$

$$11x = 3300$$

or
$$x = 300.$$

Next,
$$p = 1500 - 3(300) = 600.$$

Thus, the equilibrium quantity is 300 and the equilibrium price is $600.

25. a. We solve the system of equations $p = cx + d$ and $p = ax + b$. Substituting the first into the second gives

$$cx + d = ax + b$$

$$(c - a)x = b - d$$

or
$$x = \frac{b - d}{c - a}.$$

Since $a < 0$ and $c > 0$, and $b > d > 0$, and $c - a \neq 0$, x is well-defined. Substituting this value of x into the second equation, we obtain

$$p = a\left(\frac{b - d}{c - a}\right) + b = \frac{ab - ad + bc - ab}{c - a} = \frac{bc - ad}{c - a}.$$

Therefore, the equilibrium quantity is $\dfrac{b - d}{c - a}$ and the equilibrium price is $\dfrac{bc - ad}{c - a}$

b. If c is increased, the denominator in the expression for x increases and so x gets smaller. At the same time, the first term in the first equation for p decreases (since a is a negative number) and so p gets larger. This analysis shows that if the unit price for producing the product is increased then the equilibrium quantity decreases while the equilibrium price increases.

c. If b is decreased, the numerator of the expression for x decreases while the denominator stays the same. Therefore x decreases. The expression for p also shows that p decreases. This analysis shows that if the (theoretical) upper bound for the unit price of a commodity is lowered, then both the equilibrium quantity and the equilibrium price drop.

27. True. $P(x) = R(x) - C(x) = sx - (cx + F) = (s - c)x - F$. Therefore, the firm is making a profit if $P(x) = (s - c)x - F > 0$, or $x > \frac{F}{s-c}$ $(s \neq c)$.

29. Solving the two equations simultaneously to find the point(s) of intersection of L_1 and L_2, we obtain
$$m_1 x + b_1 = m_2 x + b_2$$
$$(m_1 - m_2)x = b_2 - b_1 \qquad (1)$$
a. If $m_1 = m_2$ and $b_2 \neq b_1$, then there is no solution for (1) and in this case L_1 and L_2 do not intersect.

b. If $m_1 \neq m_2$, then Equation (1) can be solved (uniquely) for x and this shows that L_1 and L_2 intersect at precisely one point.

c. If $m_1 = m_2$ and $b_1 = b_2$, then (1) is satisfied for all values of x and this shows that L_1 and L_2 intersect at infinitely many points.

USING TECHNOLOGY EXERCISES 1.4, page 53

1. (0.6, 6.2)   3. (3.8261, 0.1304)   5. (386.9091, 145.3939)

7.

   a.   b. (3548, 27,997)

   c. x-intercept is approximately 3548

9.   a. $C_1(x) = 34 + 0.18x;\ C_2(x) = 28 + 0.22x$

   b.   c. (150, 61)

d. If the distance driven is less than or equal to 150 mi, rent from Acme Truck Leasing; if the distance driven is more than 150 mi, rent from Ace Truck Leasing.

11. a. $p = -\dfrac{1}{10}x + 284; \quad p = \dfrac{1}{60}x + 60$

   b.   (1920, 92)

   c. 1920/wk; \$92/radio

EXERCISES 1.5, page 60

1. a. We first summarize the data:

x	y	x^2	xy
1	4	1	4
2	6	4	12
3	8	9	24
4	11	16	44
10	29	30	84

The normal equations are $4b + 10m = 29$
$$10b + 30m = 84.$$
Solving this system of equations, we obtain $m = 2.3$ and $b = 1.5$. So an equation is $y = 2.3x + 1.5$.

b. The scatter diagram and the least squares line for this data follow:

3. a. We first summarize the data:

x	y	x^2	xy
1	4.5	1	4.5
2	5	4	10
3	3	9	9
4	2	16	8
4	3.5	16	14
6	1	36	6
20	19	82	51.5

The normal equations are $6b + 20m = 19$

$$20b + 82m = 51.5.$$

The solutions are $m \approx -0.7717$ and $b \approx 5.7391$ and so a required equation is $y = -0.772x + 5.739$.

b. The scatter diagram and the least-squares line for these data follow.

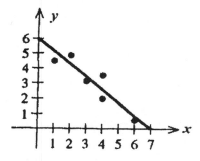

5. a. We first summarize the data:

x	y	x^2	xy
1	3	1	3
2	5	4	10
3	5	9	15
4	7	16	28
5	8	25	40
15	28	55	96

The normal equations are $55m + 15b = 96$
$$15m + 5b = 28.$$
Solving, we find $m = 1.2$ and $b = 2$, so that the required equation is $y = 1.2x + 2$.
  b. The scatter diagram and the least-squares line for the given data follow.

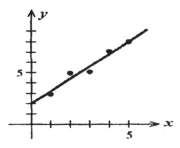

7. a. We first summarize the data:

x	y	x^2	xy
4	0.5	16	2
4.5	0.6	20.25	2.7
5	0.8	25	4
5.5	0.9	30.25	4.95
6	1.2	36	7.2
25	4	127.5	20.85

The normal equations are $\quad 5b + \quad 25m = 4$
$$25b + 127.5m = 20.85.$$
The solutions are $m = 0.34$ and $b = -0.9$, and so a required equation is
$y = 0.34x - 0.9$.

b. The scatter diagram and the least-squares line for these data follow.

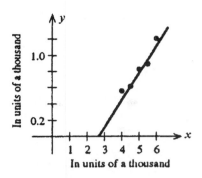

c. If $x = 6.4$, then $y = 0.34(6.4) - 0.9 = 1.276$ and so 1276 completed applications might be expected.

9. a. We first summarize the data:

x	y	x^2	xy
1	436	1	436
2	438	4	876
3	428	9	1284
4	430	16	1720
5	426	25	2130
15	2158	55	6446

The normal equations are
$$5b + 15m = 2158$$
$$15b + 55m = 6446.$$
Solving this system, we find $m = -2.8$ and $b = 440$.
Thus, the equation of the least-squares line is $y = -2.8x + 440$.
b. The scatter diagram and the least-squares line for this data are shown in the figure that follows.

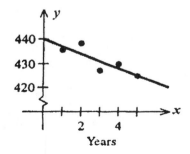

c. Two years from now, the average SAT verbal score in that area will be
$y = -2.8(7) + 440 = 420.4$.

11. a. We first summarize the data:

x	y	x^2	xy
0	168	0	0
10	213	100	2130
20	297	400	5940
30	374	900	11220
40	427	1600	17080
57	471	3249	26847
157	1950	6249	63217

The normal equations are

$$6b + 157m = 1950$$
$$157b + 6249m = 63217$$

The solutions are $m = 5.695 \approx 5.70$ and $b = 175.98 \approx 176$ and so a required equation is $y = 5.70x + 176$.

b. In 1985, $x = 45$, $y = 5.70 (45) + 176 = 432.5 \approx 433$ acres. Hence, the expected size of the average farm was 433 acres.

13. a. We first summarize the data:

x	y	x^2	xy
1	20	1	20
2	24	4	48
3	26	9	78
4	28	16	112
5	32	25	160
15	130	55	418

The normal equations are $5b + 15m = 130$
$$15b + 55m = 418.$$

The solutions are $m = 2.8$ and $b = 17.6$, and so an equation of the line is
$$y = 2.8x + 17.6.$$

b. When $x = 8$, $y = 2.8(8) + 17.6 = 40$. Hence, the state subsidy is expected to be $40 million for the eighth year.

15. a. We first summarize the data:

x	y	x^2	xy
1	16.7	1	16.7
3	26	9	78
5	33.3	25	166.5
7	48.3	49	338.1
9	57	81	513
11	65.8	121	723.8
13	74.2	169	964.6
15	83.3	225	1249.5
64	404.6	680	4050.2

The normal equations are $8b + 64m = 404.6$

$$64b + 680m = 4050.2.$$

The solutions are $m = 4.8417$ and $b = 11.8417$ and so a required equation is $y = 4.842x + 11.842$.

b. In 1993, $x = 19$, and so $y = 4.842(19) + 11.842 = 103.84$. Hence the estimated number of cans produced in 1993 is 103.8 billion.

17. a. We first summarize the data:

x	y	x^2	xy
0	21.7	0	0
1	32.1	1	32.1
2	45.0	4	90
3	58.3	9	174.9
4	69.6	16	278.4
10	226.7	30	575.4

The normal equations are $\begin{aligned} 5b + 10m &= 226.7 \\ 10b + 30m &= 575.4 \end{aligned}$.

The solutions are $m = 12.2$ and $b = 20.9$ and so a required equation is $y = 12.2x + 20.9$.

b. In 2003, $x = 5$ so $y = 12.2(5) + 20.9 = 81.9$, or 81.9 million computers are expected to be connected to the internet in Europe in that year.

19. a. We first summarize the data.

x	y	x^2	xy
0	19.5	0	0
10	20	100	200
20	20.6	400	412
30	21.2	900	636
40	21.8	1600	872
50	22.4	2500	1120
150	125.5	5500	3240

The normal equations are
$$6b + 150m = 125.5$$
$$150b + 5500m = 3240$$
The solutions are $b = 19.45$ and $m = 0.0586$. Therefore, $y = 0.059x + 19.5$.

b. The life expectancy at 65 of a female in 2040 is

$$y = 0.059(40) + 19.5 = 21.86 \quad \text{or } 21.9 \text{ years.}$$

The datum gives a life expectancy of 21.8 years.

c. The life expectancy at 65 of a female in 2030 is
$$y = 0.059(30) + 19.5 = 21.27 \quad \text{or } 21.3 \text{ years.}$$

The datum gives a life expectancy of 21.2 years.

21. a. We first summarize the given data:

x	y	x^2	xy
0	90.4	0	0
1	100.0	1	100
2	110.4	4	220.8
3	120.4	9	361.2
4	130.8	16	523.2
5	140.4	25	702
6	150	36	900
21	842.4	91	2807.2

The normal equations are
$$7b + 21m = 842.4$$
$$21b + 91m = 2807.2$$
The solutions are $m = 10$ and $b = 90.34$. Therefore, the required equation is $y = 10x + 90.34$.

b. If $x = 6$, then $y = 10(6) + 90.34 = 150.34$, or 150,340,000. This compares well with the actual data for that year-- 150,000,000 subscribers.

23. False. See Example 1, page 57.

25. True. Since there exists one and only one line passing through two distinct points, the two lines must be the same.

USING TECHNOLOGY EXERCISES 1.5, page 67

1. $y = 2.3596x + 3.8639$ 3. $y = -1.1948x + 3.5525$

5. a. $y = 0.45x + 1.28$ b. $3.08 billion

7. a. $y = 13.321x + 72.57$ b. 192 million tons

9. a. $1.95x + 12.19$; $23.89 billion   11. a. $y = 3.76x + 87.46$ b. $140,100

CHAPTER 1 REVIEW EXERCISES, page 69

1. The distance is $d = \sqrt{(6-2)^2 + (4-1)^2} = \sqrt{4^2 + 3^2} = \sqrt{25} = 5$.

3. The distance is
$$d = \sqrt{[1-(-2)]^2 + [-7-(-3)]^2} = \sqrt{3^2 + (-4)^2} = \sqrt{9+16} = \sqrt{25} = 5.$$

5. An equation is $x = -2$.

7. The slope of L is $m = \dfrac{\frac{7}{2} - 4}{3 - (-2)} = \dfrac{\frac{7-8}{2}}{5} = -\dfrac{1}{10}$ and an equation of L is
$$y - 4 = -\tfrac{1}{10}[x - (-2)] = -\tfrac{1}{10}x - \tfrac{1}{5},$$
or $y = -\tfrac{1}{10}x + \tfrac{19}{5}$

The general form of this equation is $x + 10y - 38 = 0$.

9. Writing the given equation in the form $y = \frac{5}{2}x - 3$, we see that the slope of the given line is 5/2. So a required equation is $y - 4 = \frac{5}{2}(x + 2)$ or $y = \frac{5}{2}x + 9$
The general form of this equation is $5x - 2y + 18 = 0$.

11. Using the slope-intercept form of the equation of a line, we have $y = -\frac{1}{2}x - 3$.

13. Rewriting the given equation in the slope-intercept form, we have $4y = -3x + 8$
or $\qquad y = -\frac{3}{4}x + 2$
and conclude that the slope of the required line is –3/4. Using the point-slope form of the equation of a line with the point (2,3) and slope –3/4, we obtain
$$y - 3 = -\frac{3}{4}(x - 2)$$
$$y = -\frac{3}{4}x + \frac{6}{4} + 3$$
$$= -\frac{3}{4}x + \frac{9}{2}.$$
The general form of this equation is $3x + 4y - 18 = 0$.

15. Rewriting the given equation in the slope-intercept form $y = \frac{2}{3}x - 8$, we see that the slope of the line with this equation is 2/3. The slope of the required line is $-3/2$. Using the point-slope form of the equation of a line with the point $(-2, -4)$ and slope $-3/2$, we have
$$y - (-4) = -\frac{3}{2}[x - (-2)]$$
or $\qquad y = -\frac{3}{2}x - 7$.
The general form of this equation is $3x + 2y + 14 = 0$.

17. Setting $x = 0$, gives $5y = 15$, or $y = 3$. Setting $y = 0$, gives $-2x = 15$, or $x = -15/2$.
The graph of the equation $-2x + 5y = 15$ follows.

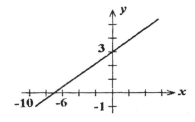

19. Let x denote the time in years. Since the function is linear, we know that it has the form $f(x) = mx + b$.

a. The slope of the line passing through (0, 2.4) and (5, 7.4) is $m = \dfrac{7.4 - 2.4}{5} = 1$.

Since the line passes through (0, 2.4), we know that the y-intercept is 2.4. Therefore, the required function is $f(x) = x + 2.4$.

b. In 1999 ($x = 3$), the sales were $f(3) = 3 + 2.4 = 5.4$, or $5.4 million dollars.

21. a. $D(w) = \dfrac{a}{150} w$. The given equation can be expressed in the form $y = mx + b$, where $m = \dfrac{a}{150}$ and $b = 0$.

b. If $a = 500$ and $w = 35$, $D(35) = \frac{500}{150}(35) = 116\frac{2}{3}$, or approximately 117 mg.

23. Let V denote the value of the machine after t years.

a. The rate of depreciation is
$$-\frac{\Delta V}{\Delta t} = \frac{300{,}000 - 30{,}000}{12} = \frac{270{,}000}{12} = 22{,}500, \quad \text{or } \$22{,}500/\text{year}.$$

b. Using the point-slope form of the equation of a line with the point (0, 300,000) and $m = -22{,}500$, we have
$$V - 300{,}000 = -22{,}500(t - 0)$$
$$V = -22{,}500t + 300{,}000.$$

25. The slope of the demand curve is $\dfrac{\Delta p}{\Delta x} = -\dfrac{10}{200} = -0.05$.

Using the point-slope form of the equation of a line with the point (0, 200), we have
$$p - 200 = -0.05(x), \quad \text{or} \quad p = -0.05x + 200.$$
The graph of the demand equation follows.

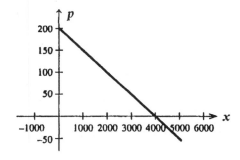

27. We solve the system $\quad 3x + 4y = -6$
$$2x + 5y = -11.$$
Solving the first equation for x, we have $3x = -4y - 6$ and
$$x = -\tfrac{4}{3}y - 2.$$
Substituting this value of x into the second equation yields
$$2(-\tfrac{4}{3}y - 2) + 5y = -11$$
$$-\tfrac{8}{3}y - 4 + 5y = -11$$
$$\tfrac{7}{3}y = -7, \qquad \text{or} \quad y = -3.$$
Then $\quad\quad\quad\quad\quad x = -\tfrac{4}{3}(-3) - 2 = 4 - 2 = 2.$
Therefore, the point of intersection is $(2, -3)$.

29. Setting $C(x) = R(x)$, we have $12x + 20,000 = 20x$
$$8x = 20,000$$
or $\quad\quad\quad\quad\quad\quad\quad\quad x = 2500.$
Next, $\quad\quad\quad\quad\quad R(2500) = 20(2500) = 50,000,$
and we conclude that the breakeven point is $(2500, 50,000)$.

31. a. The slope of the line is $m = \dfrac{1 - 0.5}{4 - 2} = 0.25.$

Using the point-slope form of an equation of a line, we have
$$y - 1 = 0.25(x - 4), \quad y = 0.25x$$
b. $\quad\quad\quad\quad y = 0.25(6.4) = 1.6, \quad$ or 1600 applications.

CHAPTER 2

EXERCISES 2.1, page 79

1. Solving the first equation for x, we find $x = 3y - 1$. Substituting this value of x into the second equation yields
    $$4(3y - 1) + 3y = 11$$
    $$12y - 4 + 3y = 11$$
    or
    $$y = 1.$$
    Substituting this value of y into the first equation gives $x = 3(1) - 1 = 2$.
    Therefore, the unique solution of the system is $(2,1)$.

3. Solving the first equation for x, we have $x = 7 - 4y$. Substituting this value of x into the second equation, we have
    $$\tfrac{1}{2}(7 - 4y) + 2y = 5$$
    $$7 - 4y + 4y = 10$$
    $$7 = 10.$$
    Clearly, this is impossible and we conclude that the system of equations has no solution.

5. Solving the first equation for x, we obtain $x = 7 - 2y$.
    Substituting this value of x into the second equation, we have
    $$2(7 - 2y) - y = 4$$
    $$14 - 4y - y = 4$$
    $$-5y = -10$$
    and
    $$y = 2.$$
    Then
    $$x = 7 - 2(2) = 7 - 4 = 3.$$
    We conclude that the solution to the system is $(3,2)$.

7. Solving the first equation for x, we have
    $$2x = 5y + 10$$
    and
    $$x = \tfrac{5}{2}y + 5.$$
    Substituting this value of x into the second equation, we have
    $$6(\tfrac{5}{2}y + 5) - 15y = 30$$
    $$15y + 30 - 15y = 30$$

or $0 = 0$.

This result tells us that the second equation is equivalent to the first. Thus, any ordered pair of numbers (x, y) satisfying the equation
$$2x - 5y = 10 \qquad (\text{or } 6x - 15y = 30)$$
is a solution to the system. In particular, by assigning the value t to x, where t is any real number, we find that $y = -2 + \frac{2}{5}t$ so the ordered pair, $(t, \frac{2}{5}t - 2)$ is a solution to the system, and we conclude that the system has infinitely many solutions.

9. Solving the first equation for x, we obtain
$$4x - 5y = 14$$
$$4x = 14 + 5y$$
$$x = \tfrac{14}{4} + \tfrac{5}{4}y = \tfrac{7}{2} + \tfrac{5}{4}y.$$
Substituting this value of x into the second equation gives
$$2(\tfrac{7}{2} + \tfrac{5}{4}y) + 3y = -4$$
$$7 + \tfrac{5}{2}y + 3y = -4$$
$$\tfrac{11}{2}y = -11$$
or $y = -2.$

Then, $x = \tfrac{7}{2} + \tfrac{5}{4}(-2) = 1.$

We conclude that the ordered pair $(1, -2)$ satisfies the given system of equations.

11. Solving the first equation for x, we obtain
$$2x = 3y + 6$$
$$x = \tfrac{3}{2}y + 3$$
Substituting this value of x into the second equation gives
$$6(\tfrac{3}{2}y + 3) - 9y = 12$$
$$9y + 18 - 9y = 12$$
$$18 = 12.$$
which is impossible. We conclude that the system of equations has no solution.

13. Solving the first equation for y, we obtain $y = 2x - 3$. Substituting this value of y into the second equation yields
$$4x + k(2x - 3) = 4,$$
$$4x + 2xk - 3k = 4$$
$$2x(2 + k) = 4 + 3k$$
$$x = \frac{4 + 3k}{2(2 + k)}.$$

Since x is not defined when the denominator of this last expression is zero, we conclude that the system has no solution when $k = -2$.

15. Let x and y denote the number of acres of corn and wheat planted, respectively. Then $x + y = 500$. Since the cost of cultivating corn is \$42/acre and that of wheat \$30/acre and Mr. Johnson has \$18,600 available for cultivation, we have
$$42x + 30y = 18600.$$
Thus, the solution is found by solving the system of equations
$$x + y = 500$$
$$42x + 30y = 18{,}600 \ .$$

17. Let x denote the number of pounds of the \$2.50/lb coffee and y denote the number of pounds of the \$3/lb coffee. Then
$$x + y = 100.$$
Since the blended coffee sells for \$2.80/lb, we know that the blended mixture is worth $(2.80)(100) = \$280$. Therefore,
$$2.50x + 3y = 280.$$
Thus, the solution is found by solving the system of equations
$$x + y = 100$$
$$2.50x + 3y = 280$$

19. Let x denote the number of children who rode the bus during the morning shift and y denote the number of adults who rode the bus during the morning shift. Then $x + y = 1000$. Since the total fare collected was \$650, we have $0.25x + 0.75y = 650$. Thus, the solution to the problem can be found by solving the system of equations
$$x + y = 1000$$
$$0.25x + 0.75y = 650.$$

21. Let x = the amount of money invested at 6 percent in a savings account
    y = the amount of money invested at 8 percent in mutual funds
and z = the amount of money invested at 12 percent in bonds.
Since the total interest was \$21,600, we have
$$0.6x + 0.8y + 0.12z = 21{,}600.$$
Also, since the amount of Mr. Sid's investment in bonds is twice the amount of the investment in the savings account, we have
$$z = 2x.$$
Finally, the interest earned from his investment in bonds was equal to the

interest earned on his money mutual funds, so
$$0.08y = 0.12z.$$

Thus, the solution to the problem can be found by solving the system of equations
$$0.06x + 0.08y + 0.12z = 21,600$$
$$2x \qquad - \quad z = \qquad 0$$
$$0.08y - 0.12z = \qquad 0.$$

23. Let x, y, and z denote the number of 100-lb bags of grade-A, grade-B, and grade-C fertilizers to be produced. The amount of nitrogen required is $18x + 20y + 24z$, and this must be equal to 26,400. So, we have $18x + 20y + 24z = 26,400$. Similarly, the constraints on the use of phosphate and potassium lead to the equations $4x + 4y + 3z = 4900$, and $5x + 4y + 6z = 6200$, respectively. Thus we have the problem of finding the solution to the system
$$18x + 20y + 24z = 26,400$$
$$4x + 4y + 3z = 4,900 \; .$$
$$5x + 4y + 6z = 6,200$$

25. Let x, y, and z denote the number of compact, intermediate, and full-size cars, respectively, to be purchased. The cost incurred in buying the specified number of cars is $10000x + 15000y + 20000z$. Since the budget is \$1.25 million, we have the system
$$10,000x + 15,000y + 20,000z = 1,250,000$$
$$x - \qquad 2y \qquad = \qquad 0$$
$$x + \qquad y + \qquad z = \qquad 100.$$

27. Let $\quad x =$ the number of ounces of Food I used in the meal
$\qquad \quad y =$ the number of ounces of Food II used in the meal
and $\quad z =$ the number of ounces of Food III used in the meal.
Since 100 percent of the daily requirement of proteins, carbohydrates, and iron is to be met by this meal, we have the following system of linear equations:
$$10x + 6y + 8z = 100$$
$$10x + 12y + 6z = 100$$
$$5x + 4y + 12z = 100.$$

29. True. If the three lines coincide, then the system has infinitely many solutions—

corresponding to all the points on the (common) line. If at least one line is distinct from the others, then the system has no solution.

EXERCISES 2.2, page 92

1. $\begin{bmatrix} 2 & -3 & | & 7 \\ 3 & 1 & | & 4 \end{bmatrix}$

3. $\begin{bmatrix} 0 & -1 & 2 & | & 6 \\ 2 & 2 & -8 & | & 7 \\ 0 & 3 & 4 & | & 0 \end{bmatrix}$

5. $3x + 2y = -4$
   $x - y = 5$

7. $x + 3y + 2z = 4$
   $2x \qquad\quad = 5$
   $3x - 3y + 2z = 6$

9. Yes. Conditions 1-4 are satisfied (see page 85 of the text).

11. No. Condition 3 is violated. The first nonzero entry in the second row does not lie to the right of the first nonzero entry 1 in row 1.

13. Yes. Conditions 1-4 are satisfied.

15. No. Condition 2 and consequently condition 4 are not satisfied. The first nonzero entry in the last row is not a 1 and the column containing that entry does not have zeros elsewhere.

17. No. Condition 1 is violated. The first row consists entirely of zeros and it lies above row 2.

19. $\begin{bmatrix} \boxed{2} & 4 & | & 8 \\ 3 & 1 & | & 2 \end{bmatrix} \xrightarrow{\frac{1}{2}R_1} \begin{bmatrix} 1 & 2 & | & 4 \\ \boxed{3} & 1 & | & 2 \end{bmatrix} \xrightarrow{R_2 - 3R_1} \begin{bmatrix} 1 & 2 & | & 4 \\ 0 & -5 & | & -10 \end{bmatrix}$

21. $\begin{bmatrix} \boxed{-1} & 2 & | & 3 \\ 6 & 4 & | & 2 \end{bmatrix} \xrightarrow{-R_1} \begin{bmatrix} 1 & -2 & | & -3 \\ 6 & 4 & | & 2 \end{bmatrix} \xrightarrow{R_2 - 6R_1} \begin{bmatrix} 1 & -2 & | & -3 \\ 0 & 16 & | & 20 \end{bmatrix}$

23.
$$\begin{bmatrix} \boxed{2} & 4 & 6 & | & 12 \\ 2 & 3 & 1 & | & 5 \\ 3 & -1 & 2 & | & 4 \end{bmatrix} \xrightarrow{\frac{1}{2}R_1} \begin{bmatrix} 1 & 2 & 3 & | & 6 \\ 2 & 3 & 1 & | & 5 \\ 3 & -1 & 2 & | & 4 \end{bmatrix} \xrightarrow[R_3-3R_1]{R_2-2R_1} \begin{bmatrix} 1 & 2 & 3 & | & 6 \\ 0 & -1 & -5 & | & -7 \\ 0 & -7 & -7 & | & -14 \end{bmatrix}$$

25.
$$\begin{bmatrix} 0 & 1 & 3 & | & 4 \\ 2 & 4 & \boxed{1} & | & 3 \\ 5 & 6 & 2 & | & -4 \end{bmatrix} \xrightarrow[R_3-2R_2]{R_1-3R_2} \begin{bmatrix} -6 & -11 & 0 & | & -5 \\ 2 & 4 & 1 & | & 3 \\ 1 & -2 & 0 & | & -10 \end{bmatrix}$$

27.
$$\begin{bmatrix} 3 & 9 & | & 6 \\ 2 & 1 & | & 4 \end{bmatrix} \xrightarrow{\frac{1}{3}R_1} \begin{bmatrix} 1 & 3 & | & 2 \\ 2 & 1 & | & 4 \end{bmatrix} \xrightarrow{R_2-2R_1} \begin{bmatrix} 1 & 3 & | & 2 \\ 0 & -5 & | & 0 \end{bmatrix} \xrightarrow{-\frac{1}{5}R_2}$$

$$\begin{bmatrix} 1 & 3 & | & 2 \\ 0 & 1 & | & 0 \end{bmatrix} \xrightarrow{R_1-3R_2} \begin{bmatrix} 1 & 0 & | & 2 \\ 0 & 1 & | & 0 \end{bmatrix}$$

29.
$$\begin{bmatrix} 1 & 3 & 1 & | & 3 \\ 3 & 8 & 3 & | & 7 \\ 2 & -3 & 1 & | & -10 \end{bmatrix} \xrightarrow[R_3-2R_1]{R_2-3R_1} \begin{bmatrix} 1 & 3 & 1 & | & 3 \\ 0 & -1 & 0 & | & -2 \\ 0 & -9 & -1 & | & -16 \end{bmatrix} \xrightarrow{-R_2} \begin{bmatrix} 1 & 3 & 1 & | & 3 \\ 0 & 1 & 0 & | & 2 \\ 0 & -9 & -1 & | & -16 \end{bmatrix}$$

$$\xrightarrow[R_3+9R_2]{R_1-3R_2} \begin{bmatrix} 1 & 0 & 1 & | & -3 \\ 0 & 1 & 0 & | & 2 \\ 0 & 0 & -1 & | & 2 \end{bmatrix} \xrightarrow[-R_3]{R_1+R_3} \begin{bmatrix} 1 & 0 & 0 & | & -1 \\ 0 & 1 & 0 & | & 2 \\ 0 & 0 & 1 & | & -2 \end{bmatrix}$$

31. The augmented matrix is equivalent to the system of linear equations
$$3x + 9y = 6$$
$$2x + y = 4$$
The ordered pair (2,0) is the solution to the system.

33. The augmented matrix is equivalent to the system of linear equations
$$x + 3y + z = 3$$
$$3x + 8y + 3z = 7$$
$$2x - 3y + z = -10$$

Reading off the solution from the last augmented matrix,

$$\left[\begin{array}{ccc|c} 1 & 0 & 0 & -1 \\ 0 & 1 & 0 & 2 \\ 0 & 0 & 1 & -2 \end{array}\right],$$

which is in row-reduced form, we have $x = -1$, $y = 2$, and $z = -2$.

35. Using the Gauss-Jordan method, we have

$$\left[\begin{array}{cc|c} 1 & -2 & 8 \\ 3 & 4 & 4 \end{array}\right] \xrightarrow{R_2 - 3R_1} \left[\begin{array}{cc|c} 1 & -2 & 8 \\ 0 & 10 & -20 \end{array}\right] \xrightarrow{\frac{1}{10}R_2} \left[\begin{array}{cc|c} 1 & -2 & 8 \\ 0 & 1 & -2 \end{array}\right] \xrightarrow{R_1 + 2R_2} \left[\begin{array}{cc|c} 1 & 0 & 4 \\ 0 & 1 & -2 \end{array}\right]$$

The solution is $(4, -2)$.

37. Using the Gauss-Jordan method, we have

$$\left[\begin{array}{cc|c} 2 & -3 & -8 \\ 4 & 1 & -2 \end{array}\right] \xrightarrow{\frac{1}{2}R_1} \left[\begin{array}{cc|c} 1 & -\frac{3}{2} & -4 \\ 4 & 1 & -2 \end{array}\right] \xrightarrow{R_2 - 4R_1} \left[\begin{array}{cc|c} 1 & -\frac{3}{2} & -4 \\ 0 & 7 & 14 \end{array}\right] \xrightarrow{\frac{1}{7}R_2}$$

$$\left[\begin{array}{cc|c} 1 & -\frac{3}{2} & -4 \\ 0 & 1 & 2 \end{array}\right] \xrightarrow{R_1 + \frac{3}{2}R_2} \left[\begin{array}{cc|c} 1 & 0 & -1 \\ 0 & 1 & 2 \end{array}\right].$$

The solution is $(-1, 2)$.

39. Using the Gauss-Jordan method, we have

$$\left[\begin{array}{ccc|c} 1 & 1 & 1 & 0 \\ 2 & -1 & 1 & 1 \\ 1 & 1 & -2 & 2 \end{array}\right] \xrightarrow[R_3 - R_1]{R_2 - 2R_1} \left[\begin{array}{ccc|c} 1 & 1 & 1 & 0 \\ 0 & -3 & -1 & 1 \\ 0 & 0 & -3 & 2 \end{array}\right] \xrightarrow{-\frac{1}{3}R_2} \left[\begin{array}{ccc|c} 1 & 1 & 1 & 0 \\ 0 & 1 & \frac{1}{3} & -\frac{1}{3} \\ 0 & 0 & -3 & 2 \end{array}\right] \xrightarrow{R_1 - R_2}$$

$$\left[\begin{array}{ccc|c} 1 & 0 & \frac{2}{3} & \frac{1}{3} \\ 0 & 1 & \frac{1}{3} & -\frac{1}{3} \\ 0 & 0 & -3 & 2 \end{array}\right] \xrightarrow{-\frac{1}{3}R_3} \left[\begin{array}{ccc|c} 1 & 0 & \frac{2}{3} & \frac{1}{3} \\ 0 & 1 & \frac{1}{3} & -\frac{1}{3} \\ 0 & 0 & 1 & -\frac{2}{3} \end{array}\right] \xrightarrow[R_2 - \frac{1}{3}R_3]{R_1 - \frac{2}{3}R_3} \left[\begin{array}{ccc|c} 1 & 0 & 0 & \frac{7}{9} \\ 0 & 1 & 0 & -\frac{1}{9} \\ 0 & 0 & 1 & -\frac{2}{3} \end{array}\right].$$

The solution is $\left(\frac{7}{9}, -\frac{1}{9}, -\frac{2}{3}\right)$.

41.
$$\begin{bmatrix} 2 & 2 & 1 & | & 9 \\ 1 & 0 & 1 & | & 4 \\ 0 & 4 & -3 & | & 17 \end{bmatrix} \xrightarrow{R_1 \leftrightarrow R_2} \begin{bmatrix} 1 & 0 & 1 & | & 4 \\ 2 & 2 & 1 & | & 9 \\ 0 & 4 & -3 & | & 17 \end{bmatrix} \xrightarrow{R_2 - 2R_1} \begin{bmatrix} 1 & 0 & 1 & | & 4 \\ 0 & 2 & -1 & | & 1 \\ 0 & 4 & -3 & | & 17 \end{bmatrix} \xrightarrow{\frac{1}{2}R_2}$$

$$\begin{bmatrix} 1 & 0 & 1 & | & 4 \\ 0 & 1 & -\frac{1}{2} & | & \frac{1}{2} \\ 0 & 4 & -3 & | & 17 \end{bmatrix} \xrightarrow{R_3 - 4R_2} \begin{bmatrix} 1 & 0 & 1 & | & 4 \\ 0 & 1 & -\frac{1}{2} & | & \frac{1}{2} \\ 0 & 0 & -1 & | & 15 \end{bmatrix} \xrightarrow{-R_3} \begin{bmatrix} 1 & 0 & 1 & | & 4 \\ 0 & 1 & -\frac{1}{2} & | & \frac{1}{2} \\ 0 & 0 & 1 & | & -15 \end{bmatrix} \xrightarrow[R_2 + \frac{1}{2}R_3]{R_1 - R_3}$$

$$\begin{bmatrix} 1 & 0 & 0 & | & 19 \\ 0 & 1 & 0 & | & -7 \\ 0 & 0 & 1 & | & -15 \end{bmatrix}. \qquad \text{The solution is } (19, -7, -15).$$

43.
$$\begin{bmatrix} 0 & -1 & 1 & | & 2 \\ 4 & -3 & 2 & | & 16 \\ 3 & 2 & 1 & | & 11 \end{bmatrix} \xrightarrow{R_1 \leftrightarrow R_2} \begin{bmatrix} 4 & -3 & 2 & | & 16 \\ 0 & -1 & 1 & | & 2 \\ 3 & 2 & 1 & | & 11 \end{bmatrix} \xrightarrow{R_1 - R_3} \begin{bmatrix} 1 & -5 & 1 & | & 5 \\ 0 & -1 & 1 & | & 2 \\ 3 & 2 & 1 & | & 11 \end{bmatrix}$$

$$\xrightarrow[R_3 - 3R_1]{-R_2} \begin{bmatrix} 1 & -5 & 1 & | & 5 \\ 0 & 1 & -1 & | & -2 \\ 0 & 17 & -2 & | & -4 \end{bmatrix} \xrightarrow[R_3 - 17R_2]{R_1 + 5R_2} \begin{bmatrix} 1 & 0 & -4 & | & -5 \\ 0 & 1 & -1 & | & -2 \\ 0 & 0 & 15 & | & 30 \end{bmatrix} \xrightarrow{\frac{1}{15}R_3}$$

$$\begin{bmatrix} 1 & 0 & -4 & | & -5 \\ 0 & 1 & -1 & | & -2 \\ 0 & 0 & 1 & | & 2 \end{bmatrix} \xrightarrow[R_2 + R_3]{R_1 + 4R_3} \begin{bmatrix} 1 & 0 & 0 & | & 3 \\ 0 & 1 & 0 & | & 0 \\ 0 & 0 & 1 & | & 2 \end{bmatrix}.$$

The solution is (3,0,2).

45. Using the Gauss-Jordan method, we have

$$\begin{bmatrix} 1 & -2 & 1 & | & 6 \\ 2 & 1 & -3 & | & -3 \\ 1 & -3 & 3 & | & 10 \end{bmatrix} \xrightarrow[R_3 - R_1]{R_2 - 2R_1} \begin{bmatrix} 1 & -2 & 1 & | & 6 \\ 0 & 5 & -5 & | & -15 \\ 0 & -1 & 2 & | & 4 \end{bmatrix} \xrightarrow{\frac{1}{5}R_2} \begin{bmatrix} 1 & -2 & 1 & | & 6 \\ 0 & 1 & -1 & | & -3 \\ 0 & -1 & 2 & | & 4 \end{bmatrix}$$

$$\xrightarrow[\substack{R_1+2R_2 \\ R_3+R_2}]{} \begin{bmatrix} 1 & 0 & -1 & 0 \\ 0 & 1 & -1 & -3 \\ 0 & 0 & 1 & 1 \end{bmatrix} \xrightarrow[\substack{R_1+R_3 \\ R_2+R_3}]{} \begin{bmatrix} 1 & 0 & 0 & 1 \\ 0 & 1 & 0 & -2 \\ 0 & 0 & 1 & 1 \end{bmatrix}.$$

Therefore, the solution is $(1,-2,1)$.

47. Using the Gauss-Jordan method, we have

$$\begin{bmatrix} 2 & 0 & 3 & -1 \\ 3 & -2 & 1 & 9 \\ 1 & 1 & 4 & 4 \end{bmatrix} \xrightarrow{R_1 \leftrightarrow R_3} \begin{bmatrix} 1 & 1 & 4 & 4 \\ 3 & -2 & 1 & 9 \\ 2 & 0 & 3 & -1 \end{bmatrix} \xrightarrow[\substack{R_2-3R_1 \\ R_3-2R_1}]{} \begin{bmatrix} 1 & 1 & 4 & 4 \\ 0 & -5 & -11 & -3 \\ 0 & -2 & -5 & -9 \end{bmatrix}$$

$$\xrightarrow{-\frac{1}{5}R_2} \begin{bmatrix} 1 & 1 & 4 & 4 \\ 0 & 1 & \frac{11}{5} & \frac{3}{5} \\ 0 & -2 & -5 & -9 \end{bmatrix} \xrightarrow[\substack{R_1-R_2 \\ R_3+2R_2}]{} \begin{bmatrix} 1 & 0 & \frac{9}{5} & \frac{17}{5} \\ 0 & 1 & \frac{11}{5} & \frac{3}{5} \\ 0 & 0 & -\frac{3}{5} & -\frac{39}{5} \end{bmatrix} \xrightarrow{-\frac{5}{3}R_3}$$

$$\begin{bmatrix} 1 & 0 & \frac{9}{5} & \frac{17}{5} \\ 0 & 1 & \frac{11}{5} & \frac{3}{5} \\ 0 & 0 & 1 & 13 \end{bmatrix} \xrightarrow[\substack{R_1-\frac{9}{5}R_3 \\ R_2-\frac{11}{5}R_3}]{} \begin{bmatrix} 1 & 0 & 0 & -20 \\ 0 & 1 & 0 & -28 \\ 0 & 0 & 1 & 13 \end{bmatrix}.$$

Therefore, the solution is $(-20, -28, 13)$.

49. Using the Gauss-Jordan method, we have

$$\begin{bmatrix} 1 & -1 & 3 & 14 \\ 1 & 1 & 1 & 6 \\ -2 & -1 & 1 & -4 \end{bmatrix} \xrightarrow[\substack{R_2-R_1 \\ R_3+2R_1}]{} \begin{bmatrix} 1 & -1 & 3 & 14 \\ 0 & 2 & -2 & -8 \\ 0 & -3 & 7 & 24 \end{bmatrix} \xrightarrow{\frac{1}{2}R_2} \begin{bmatrix} 1 & -1 & 3 & 14 \\ 0 & 1 & -1 & -4 \\ 0 & -3 & 7 & 24 \end{bmatrix}$$

$$\xrightarrow[\substack{R_1+R_2 \\ R_3+3R_2}]{} \begin{bmatrix} 1 & 0 & 2 & 10 \\ 0 & 1 & -1 & -4 \\ 0 & 0 & 4 & 12 \end{bmatrix} \xrightarrow{\frac{1}{4}R_3} \begin{bmatrix} 1 & 0 & 2 & 10 \\ 0 & 1 & -1 & -4 \\ 0 & 0 & 1 & 3 \end{bmatrix} \xrightarrow[\substack{R_1-2R_3 \\ R_2+R_3}]{} \begin{bmatrix} 1 & 0 & 0 & 4 \\ 0 & 1 & 0 & -1 \\ 0 & 0 & 1 & 3 \end{bmatrix}$$

Therefore, the solution is $(4, -1, 3)$.

51. We wish to solve the system of equations

$$x + y = 500 \qquad (x = \text{the number of acres of corn planted})$$
$$42x + 30y = 18{,}600 \qquad (y = \text{the number of acres of wheat planted})$$

Using the Gauss-Jordan method, we find

$$\begin{bmatrix} 1 & 1 & | & 500 \\ 42 & 30 & | & 18600 \end{bmatrix} \xrightarrow{R_2-42R_1} \begin{bmatrix} 1 & 1 & | & 500 \\ 0 & -12 & | & -2400 \end{bmatrix} \xrightarrow{-\frac{1}{12}R_2} \begin{bmatrix} 1 & 1 & | & 500 \\ 0 & 1 & | & 200 \end{bmatrix}$$

$$\xrightarrow{R_1-R_2} \begin{bmatrix} 1 & 0 & | & 300 \\ 0 & 1 & | & 200 \end{bmatrix}.$$

The solution to this system of equations is $x = 300$ and $y = 200$. We conclude that Jacob should plant 300 acres of corn and 200 acres of wheat.

53. Let x denote the number of pounds of the \$2.50/lb coffee and y denote the number of pounds of the \$3.00/lb coffee. Then we are required to solve the system
$$x + y = 100$$
$$2.50x + 3.00y = 280$$
Using the Gauss-Jordan method of elimination, we have

$$\begin{bmatrix} 1 & 1 & | & 100 \\ 2.5 & 3 & | & 280 \end{bmatrix} \xrightarrow{R_2-2.5R_1} \begin{bmatrix} 1 & 1 & | & 100 \\ 0 & 0.5 & | & 30 \end{bmatrix} \xrightarrow{2R_2} \begin{bmatrix} 1 & 1 & | & 100 \\ 0 & 1 & | & 60 \end{bmatrix}$$

$$\xrightarrow{R_1-R_2} \begin{bmatrix} 1 & 0 & | & 40 \\ 0 & 1 & | & 60 \end{bmatrix}.$$

Therefore, 40 pounds of the \$2.50/lb coffee and 60 pounds of the \$3.00/lb coffee should be used in the 100 lb mixture.

55. Let x and y denote the number of children and adults who rode the bus during the morning shift, respectively. Then the solution to the problem can be found by solving the system of equations
$$x + y = 1000$$
$$0.25x + 0.75y = 650$$
Using the Gauss-Jordan elimination method, we have

$$\begin{bmatrix} 1 & 1 & | & 1000 \\ 0.25 & 0.75 & | & 650 \end{bmatrix} \xrightarrow{R_2-0.25R_1} \begin{bmatrix} 1 & 1 & | & 1000 \\ 0 & 0.5 & | & 400 \end{bmatrix} \xrightarrow{2R_2} \begin{bmatrix} 1 & 1 & | & 1000 \\ 0 & 1 & | & 800 \end{bmatrix}$$

$$\xrightarrow{R_1 - R_2} \begin{bmatrix} 1 & 0 & | & 200 \\ 0 & 1 & | & 800 \end{bmatrix}.$$

We conclude that 800 adults and 200 children rode the bus during the morning shift.

57. Let x, y, and z, denote the amount of money he should invest in a savings account, in mutual funds, and in bonds, respectively. Then, we are required to solve the system

$$0.06x + 0.08y + 0.12z = 21{,}600$$
$$2x \quad\quad - \quad z = 0$$
$$0.08y - 0.12z = 0$$

Using the Gauss-Jordan method, we find

$$\begin{bmatrix} 0.06 & 0.08 & 0.12 & | & 21{,}600 \\ 2 & 0 & -1 & | & 0 \\ 0 & 0.08 & -0.12 & | & 0 \end{bmatrix} \xrightarrow[\frac{1}{0.08}R_3]{\frac{1}{0.06}R_1} \begin{bmatrix} 1 & \frac{4}{3} & 2 & | & 360{,}000 \\ 2 & 0 & -1 & | & 0 \\ 0 & 1 & -\frac{3}{2} & | & 0 \end{bmatrix} \xrightarrow{R_2 - 2R_1}$$

$$\begin{bmatrix} 1 & \frac{4}{3} & 2 & | & 360{,}000 \\ 0 & -\frac{8}{3} & -5 & | & -720{,}000 \\ 0 & 1 & -\frac{3}{2} & | & 0 \end{bmatrix} \xrightarrow{-\frac{3}{8}R_2} \begin{bmatrix} 1 & \frac{4}{3} & 2 & | & 360{,}000 \\ 0 & 1 & \frac{15}{8} & | & 270{,}000 \\ 0 & 1 & -\frac{3}{2} & | & 0 \end{bmatrix}$$

$$\xrightarrow[R_3 - R_2]{R_1 - \frac{4}{3}R_2} \begin{bmatrix} 1 & 0 & -\frac{1}{2} & | & 0 \\ 0 & 1 & \frac{15}{8} & | & 270{,}000 \\ 0 & 0 & -\frac{27}{8} & | & -270{,}000 \end{bmatrix} \xrightarrow{-\frac{8}{27}R_3} \begin{bmatrix} 1 & 0 & -\frac{1}{2} & | & 0 \\ 0 & 1 & \frac{15}{8} & | & 270{,}000 \\ 0 & 0 & 1 & | & 80{,}000 \end{bmatrix}$$

$$\xrightarrow[R_2 - \frac{15}{8}R_3]{R_1 + \frac{1}{2}R_3} \begin{bmatrix} 1 & 0 & 0 & | & 40{,}000 \\ 0 & 1 & 0 & | & 120{,}000 \\ 0 & 0 & 1 & | & 80{,}000 \end{bmatrix}$$

Therefore, Sid should invest $40,000 in a savings account, $120,000 in mutual funds, and $80,000 in bonds.

59. Refer to Exercise 23, page 75. We obtain the following augmented matrices.

$$\begin{bmatrix} 18 & 20 & 24 & 26400 \\ 4 & 4 & 3 & 4900 \\ 5 & 4 & 6 & 6200 \end{bmatrix} \xrightarrow{R_1 \leftrightarrow R_3} \begin{bmatrix} 5 & 4 & 6 & 6200 \\ 4 & 4 & 3 & 4900 \\ 18 & 20 & 24 & 26400 \end{bmatrix} \xrightarrow{R_1 - R_2}$$

$$\begin{bmatrix} 1 & 0 & 3 & 1300 \\ 4 & 4 & 3 & 4900 \\ 18 & 20 & 24 & 26400 \end{bmatrix} \xrightarrow[R_3 - 18R_1]{R_2 - 4R_1} \begin{bmatrix} 1 & 0 & 3 & 1300 \\ 0 & 4 & -9 & -300 \\ 0 & 20 & -30 & 3000 \end{bmatrix} \xrightarrow{\frac{1}{4}R_2}$$

$$\begin{bmatrix} 1 & 0 & 3 & 1300 \\ 0 & 1 & -\frac{9}{4} & -75 \\ 0 & 20 & -30 & 3000 \end{bmatrix} \xrightarrow{R_3 - 20R_2} \begin{bmatrix} 1 & 0 & 3 & 1300 \\ 0 & 1 & -\frac{9}{4} & -75 \\ 0 & 0 & 15 & 4500 \end{bmatrix} \xrightarrow{\frac{1}{15}R_3}$$

$$\begin{bmatrix} 1 & 0 & 3 & 1300 \\ 0 & 1 & -\frac{9}{4} & -75 \\ 0 & 0 & 1 & 300 \end{bmatrix} \xrightarrow[R_2 + \frac{9}{4}R_3]{R_1 - 3R_3} \begin{bmatrix} 1 & 0 & 0 & 400 \\ 0 & 1 & 0 & 600 \\ 0 & 0 & 1 & 300 \end{bmatrix}$$

We see that $x = 400$, $y = 600$, and $z = 300$. Therefore, Lawnco should produce 400, 600, and 300 100-lb bags of grade-A, grade-B, and grade-C fertilizer.

61. Let x, y, and z denote the number of compact, intermediate, and full-size cars, respectively, to be purchased. Then the problem can be solved by solving the system

$$10000x + 15,000y + 20,000z = 1,250,000$$
$$x - 2y = 0$$
$$x + y + z = 100$$

Using the Gauss-Jordan method, we have

$$\begin{bmatrix} 10,000 & 15,000 & 20,000 & 1,250,000 \\ 1 & -2 & 0 & 0 \\ 1 & 1 & 1 & 100 \end{bmatrix} \xrightarrow{R_1 \leftrightarrow R_3} \begin{bmatrix} 1 & 1 & 1 & 100 \\ 1 & -2 & 0 & 0 \\ 10000 & 15000 & 20000 & 1,250,000 \end{bmatrix}$$

$$\xrightarrow[R_3 - 10,000R_1]{R_2 - R_1} \begin{bmatrix} 1 & 1 & 1 & 100 \\ 0 & -3 & -1 & -100 \\ 0 & 5000 & 10000 & 250,000 \end{bmatrix} \xrightarrow{-\frac{1}{3}R_2} \begin{bmatrix} 1 & 1 & 1 & 100 \\ 0 & 1 & \frac{1}{3} & \frac{100}{3} \\ 0 & 5000 & 10000 & 250,000 \end{bmatrix}$$

$$\begin{array}{c} R_1-R_2 \\ \underline{\quad R_3-5000R_2\quad} \end{array} \rightarrow \begin{bmatrix} 1 & 0 & \frac{2}{3} & \frac{200}{3} \\ 0 & 1 & \frac{1}{3} & \frac{100}{3} \\ 0 & 0 & \frac{25{,}000}{3} & \frac{250{,}000}{3} \end{bmatrix} \xrightarrow{\frac{3}{25{,}000}R_3} \begin{bmatrix} 1 & 0 & \frac{2}{3} & \frac{200}{3} \\ 0 & 1 & \frac{1}{3} & \frac{100}{3} \\ 0 & 0 & 1 & 10 \end{bmatrix} \begin{array}{c} R_1-\frac{2}{3}R_3 \\ \underline{\quad R_2-\frac{1}{3}R_3\quad} \end{array} \rightarrow \begin{bmatrix} 1 & 0 & 0 & 60 \\ 0 & 1 & 0 & 30 \\ 0 & 0 & 1 & 10 \end{bmatrix}.$$

We conclude that 60 compact cars, 30 intermediate-size cars, and 10 full-size cars will be purchased.

63. Let x, y, and z, represent the number of ounces of Food *I*, Food *II*, and Food *III* used in the meal, respectively. Then the problem reduces to solving the following system of linear equations:

$$10x + 6y + 8z = 100$$
$$10x + 12y + 6z = 100$$
$$5x + 4y + 12z = 100.$$

Using the Gauss-Jordan method, we obtain

$$\begin{bmatrix} 10 & 6 & 8 & 100 \\ 10 & 12 & 6 & 100 \\ 5 & 4 & 12 & 100 \end{bmatrix} \xrightarrow{\frac{1}{10}R_1} \begin{bmatrix} 1 & \frac{3}{5} & \frac{4}{5} & 10 \\ 10 & 12 & 6 & 100 \\ 5 & 4 & 12 & 100 \end{bmatrix} \begin{array}{c} R_2-10R_1 \\ \underline{\quad R_3-5R_1\quad} \end{array} \rightarrow$$

$$\begin{bmatrix} 1 & \frac{3}{5} & \frac{4}{5} & 10 \\ 0 & 6 & -2 & 0 \\ 0 & 1 & 8 & 50 \end{bmatrix} \xrightarrow{\frac{1}{6}R_2} \begin{bmatrix} 1 & \frac{3}{5} & \frac{4}{5} & 10 \\ 0 & 1 & -\frac{1}{3} & 0 \\ 0 & 1 & 8 & 50 \end{bmatrix} \begin{array}{c} R_1-\frac{3}{5}R_2 \\ \underline{\quad R_3-R_2\quad} \end{array} \rightarrow$$

$$\begin{bmatrix} 1 & 0 & 1 & 10 \\ 0 & 1 & -\frac{1}{3} & 0 \\ 0 & 0 & \frac{25}{3} & 50 \end{bmatrix} \xrightarrow{\frac{3}{25}R_3} \begin{bmatrix} 1 & 0 & 1 & 10 \\ 0 & 1 & -\frac{1}{3} & 0 \\ 0 & 0 & 1 & 6 \end{bmatrix} \begin{array}{c} R_1-R_3 \\ \underline{\quad R_2+\frac{1}{3}R_3\quad} \end{array} \rightarrow \begin{bmatrix} 1 & 0 & 0 & 4 \\ 0 & 1 & 0 & 2 \\ 0 & 0 & 1 & 6 \end{bmatrix}.$$

We conclude that 4 oz of Food *I*, 2 oz of Food *II*, and 6 oz of Food *III* should be used to prepare the meal.

65. Let $x =$ the number of front orchestra seats sold
            $y =$ the number of rear orchestra seats sold
and $z =$ the number of front balcony seats sold for this performance.
Then, we are required to solve the system

$$x + y + z = 1{,}000$$
$$80x + 60y + 50z = 62{,}800$$
$$x + y - 2z = 400.$$

Using the Gauss-Jordan method, we find

$$
\begin{bmatrix}
1 & 1 & 1 & 1{,}000 \\
80 & 60 & 50 & 62{,}800 \\
1 & 1 & -2 & 400
\end{bmatrix}
\xrightarrow[\substack{R_2-80R_1 \\ R_3-R_1}]{}
\begin{bmatrix}
1 & 1 & 1 & 1{,}000 \\
0 & -20 & -30 & -17{,}200 \\
0 & 0 & -3 & -600
\end{bmatrix}
\xrightarrow[\substack{-\frac{1}{20}R_2 \\ -\frac{1}{3}R_3}]{}
$$

$$
\begin{bmatrix}
1 & 1 & 1 & 1{,}000 \\
0 & 1 & \frac{3}{2} & 860 \\
0 & 0 & 1 & 200
\end{bmatrix}
\xrightarrow{R_1-R_2}
\begin{bmatrix}
1 & 0 & -\frac{1}{2} & 140 \\
0 & 1 & \frac{3}{2} & 860 \\
0 & 0 & 1 & 200
\end{bmatrix}
\xrightarrow[\substack{R_1+\frac{1}{2}R_3 \\ R_2-\frac{3}{2}R_3}]{}
$$

$$
\begin{bmatrix}
1 & 0 & 0 & 240 \\
0 & 1 & 0 & 560 \\
0 & 0 & 1 & 200
\end{bmatrix}.
$$

We conclude that tickets for 240 front orchestra seats, 560 rear orchestra seats, and 200 front balcony seats were sold.

67. Let x, y, and z denote the number of days he spent in London, Paris, and Rome, respectively. We have
$$180x + 230y + 160z = 2660$$
$$110x + 120y + 90z = 1520$$
$$x - y - z = 0 \qquad \text{(since } x = y + z\text{)}$$
Using the Gauss-Joran method to solve the system, we have

$$
\begin{bmatrix}
180 & 230 & 160 & 2660 \\
110 & 120 & 90 & 1520 \\
1 & -1 & -1 & 0
\end{bmatrix}
\xrightarrow{R_1 \leftrightarrow R_3}
\begin{bmatrix}
1 & -1 & -1 & 0 \\
110 & 120 & 90 & 1520 \\
180 & 230 & 160 & 2660
\end{bmatrix}
\xrightarrow[\substack{R_2-110R_1 \\ R_3-180R_1}]{}
$$

$$
\begin{bmatrix}
1 & -1 & -1 & 0 \\
0 & 230 & 200 & 1520 \\
0 & 410 & 340 & 2660
\end{bmatrix}
\xrightarrow{\frac{1}{230}R_2}
\begin{bmatrix}
1 & -1 & -1 & 0 \\
0 & 1 & \frac{20}{23} & \frac{152}{23} \\
0 & 410 & 340 & 2660
\end{bmatrix}
\xrightarrow[\substack{R_1+R_2 \\ R_3-410R_2}]{}
$$

$$\begin{bmatrix} 1 & 0 & -\frac{3}{23} & \frac{152}{23} \\ 0 & 1 & \frac{20}{23} & \frac{152}{23} \\ 0 & 0 & -\frac{380}{23} & -\frac{1140}{23} \end{bmatrix} \xrightarrow{-\frac{23}{380}R_3} \begin{bmatrix} 1 & 0 & -\frac{3}{23} & \frac{152}{23} \\ 0 & 1 & \frac{20}{23} & \frac{152}{23} \\ 0 & 0 & 1 & 3 \end{bmatrix} \xrightarrow[R_2 - \frac{20}{23}R_1]{R_1 + \frac{3}{23}R_3}$$

$$\begin{bmatrix} 1 & 0 & 0 & 7 \\ 0 & 1 & 0 & 4 \\ 0 & 0 & 1 & 3 \end{bmatrix}$$

The solution is $x = 7$, y = 4, and $z = 3$. Therefore, he spent 7 days in London, 4 days in Paris, and 3 days in Rome.

69. False. The constant cannot be zero. The system
$$2x + y = 1$$
$$3x - y = 2$$
is not equivalent to
$$\begin{aligned} 2x + y &= 1 \\ 0(3x - y) &= 0(2) \end{aligned} \quad \text{or} \quad \begin{aligned} 2x + y &= 1 \\ 0 &= 0 \end{aligned}.$$

USING TECHNOLOGY EXERCISES 2.2, page 96

1. $(3,1,-1,2)$ 　　　　　 3. $(5,4,-3,-4)$ 　　　　　 5. $(1,-1,2,0,3)$

EXERCISES 2.3, page 105

1. a. The system has one solution. 　　 b. The solution is $(3, -1, 2)$.

3. a. The system has one solution. 　　 b. The solution is $(2, 4)$.

5. a. The system has infinitely many solutions.
   b. Letting $x_3 = t$, we see that the solutions are given by $(4 - t, -2, t)$, where t is a parameter.

7. a. The system has no solution. The last row contains all zeros to the left of the vertical line and a nonzero number (1) to its right.

9. a. The system has infinitely many solutions.
   b. Letting $x_4 = t$, we see that the solutions are given by $(2, -1, 2 - t, t)$, where t is a parameter.

11. a. The system has infinitely many solutions.
    b. Letting $x_3 = s$ and $x_4 = t$, the solutions are given by $(2 - 3s, 1 + s, s, t)$, where s and t are parameters.

13. Using the Gauss-Jordan method, we have

$$\begin{bmatrix} 2 & -1 & 3 \\ 1 & 2 & 4 \\ 2 & 3 & 7 \end{bmatrix} \xrightarrow{R_1 \leftrightarrow R_2} \begin{bmatrix} 1 & 2 & 4 \\ 2 & -1 & 3 \\ 2 & 3 & 7 \end{bmatrix} \xrightarrow[R_3 - 2R_1]{R_2 - 2R_1} \begin{bmatrix} 1 & 2 & 4 \\ 0 & -5 & -5 \\ 0 & -1 & -1 \end{bmatrix} \xrightarrow{-\frac{1}{5}R_2}$$

$$\begin{bmatrix} 1 & 2 & 4 \\ 0 & 1 & 1 \\ 0 & -1 & -1 \end{bmatrix} \xrightarrow[R_3 + R_2]{R_1 - 2R_2} \begin{bmatrix} 1 & 0 & 2 \\ 0 & 1 & 1 \\ 0 & 0 & 0 \end{bmatrix}. \qquad \text{The solution is } (2,1).$$

15. Using the Gauss-Jordan method, we have

$$\begin{bmatrix} 3 & -2 & -3 \\ 2 & 1 & 3 \\ 1 & -2 & -5 \end{bmatrix} \xrightarrow{R_1 \leftrightarrow R_3} \begin{bmatrix} 1 & -2 & -5 \\ 2 & 1 & 3 \\ 3 & -2 & -3 \end{bmatrix} \xrightarrow[R_3 - 3R_1]{R_2 - 2R_1} \begin{bmatrix} 1 & -2 & -5 \\ 0 & 5 & 13 \\ 0 & 4 & 12 \end{bmatrix} \xrightarrow{\frac{1}{5}R_2}$$

$$\begin{bmatrix} 1 & -2 & -5 \\ 0 & 1 & \frac{13}{5} \\ 0 & 4 & 12 \end{bmatrix} \xrightarrow[R_3 - 4R_2]{R_1 + 2R_2} \begin{bmatrix} 1 & 0 & \frac{1}{5} \\ 0 & 1 & \frac{13}{5} \\ 0 & 0 & \frac{8}{5} \end{bmatrix}.$$

Since the last row implies the $0 = 8/5$, we conclude that the system of equations is inconsistent and has no solution.

17. $$\begin{bmatrix} 3 & -2 & 5 \\ -1 & 3 & -4 \\ 2 & -4 & 6 \end{bmatrix} \xrightarrow{R_1 \leftrightarrow R_2} \begin{bmatrix} -1 & 3 & -4 \\ 3 & -2 & 5 \\ 2 & -4 & 6 \end{bmatrix} \xrightarrow{-R_1} \begin{bmatrix} 1 & -3 & 4 \\ 3 & -2 & 5 \\ 2 & -4 & 6 \end{bmatrix} \xrightarrow[R_3 - 2R_1]{R_2 - 3R_1}$$

$$\begin{bmatrix} 1 & -3 & | & 4 \\ 0 & 7 & | & -7 \\ 0 & 2 & | & -2 \end{bmatrix} \xrightarrow{\frac{1}{7}R_2} \begin{bmatrix} 1 & -3 & | & 4 \\ 0 & 1 & | & -1 \\ 0 & 2 & | & -2 \end{bmatrix} \xrightarrow[R_3-2R_2]{R_1+3R_2} \begin{bmatrix} 1 & 0 & | & 1 \\ 0 & 1 & | & -1 \\ 0 & 0 & | & 0 \end{bmatrix}.$$

We conclude that the solution is $(1,-1)$.

19. $$\begin{bmatrix} 1 & -2 & | & 2 \\ 7 & -14 & | & 14 \\ 3 & -6 & | & 6 \end{bmatrix} \xrightarrow[R_3-3R_1]{R_2-7R_1} \begin{bmatrix} 1 & -2 & | & 2 \\ 0 & 0 & | & 0 \\ 0 & 0 & | & 0 \end{bmatrix}.$$

We conclude that the infinitely many solutions are given by $(2t+2, t)$, where t is a parameter.

21. $$\begin{bmatrix} 3 & 2 & | & 4 \\ -\frac{3}{2} & -1 & | & -2 \\ 6 & 4 & | & 8 \end{bmatrix} \xrightarrow{\frac{1}{3}R_1} \begin{bmatrix} 1 & \frac{2}{3} & | & \frac{4}{3} \\ -\frac{3}{2} & -1 & | & -2 \\ 6 & 4 & | & 8 \end{bmatrix} \xrightarrow[R_3-6R_1]{R_2+\frac{3}{2}R_1} \begin{bmatrix} 1 & \frac{2}{3} & | & \frac{4}{3} \\ 0 & 0 & | & 0 \\ 0 & 0 & | & 0 \end{bmatrix}.$$

We conclude that the infinitely many solutions are given by $(\frac{4}{3} - \frac{2}{3}t, t)$, where t is a parameter.

23. $$\begin{bmatrix} 2 & -1 & 1 & | & -4 \\ 3 & -\frac{3}{2} & \frac{3}{2} & | & -6 \\ -6 & 3 & -3 & | & 12 \end{bmatrix} \xrightarrow{\frac{1}{2}R_1} \begin{bmatrix} 1 & -\frac{1}{2} & \frac{1}{2} & | & -2 \\ 3 & -\frac{3}{2} & \frac{3}{2} & | & -6 \\ -6 & 3 & -3 & | & 12 \end{bmatrix} \xrightarrow[R_3+6R_1]{R_2-3R_1} \begin{bmatrix} 1 & -\frac{1}{2} & \frac{1}{2} & | & -2 \\ 0 & 0 & 0 & | & 0 \\ 0 & 0 & 0 & | & 0 \end{bmatrix}.$$

We conclude that the infinitely many solutions are given by $(-2+\frac{1}{2}s-\frac{1}{2}t, s, t)$ where s and t are parameters.

25. $$\begin{bmatrix} 1 & -2 & 3 & | & 4 \\ 2 & 3 & -1 & | & 2 \\ 1 & 2 & -3 & | & -6 \end{bmatrix} \xrightarrow[R_3-R_1]{R_2-2R_1} \begin{bmatrix} 1 & -2 & 3 & | & 4 \\ 0 & 7 & -7 & | & -6 \\ 0 & 4 & -6 & | & -10 \end{bmatrix} \xrightarrow{\frac{1}{7}R_2} \begin{bmatrix} 1 & -2 & 3 & | & 4 \\ 0 & 1 & -1 & | & -\frac{6}{7} \\ 0 & 4 & -6 & | & -10 \end{bmatrix}$$

$$\xrightarrow[R_3-4R_2]{R_1+2R_2} \begin{bmatrix} 1 & 0 & 1 & | & \frac{16}{7} \\ 0 & 1 & -1 & | & -\frac{6}{7} \\ 0 & 0 & -2 & | & -\frac{46}{7} \end{bmatrix} \xrightarrow{-\frac{1}{2}R_3} \begin{bmatrix} 1 & 0 & 1 & | & \frac{16}{7} \\ 0 & 1 & -1 & | & -\frac{6}{7} \\ 0 & 0 & 1 & | & \frac{23}{7} \end{bmatrix} \xrightarrow[R_2+R_3]{R_1-R_3}$$

$$\begin{bmatrix} 1 & 0 & 0 & | & -1 \\ 0 & 1 & 0 & | & \frac{17}{7} \\ 0 & 0 & 1 & | & \frac{23}{7} \end{bmatrix}.$$

We conclude that the solution is $(-1, \frac{17}{7}, \frac{23}{7})$.

27. $\begin{bmatrix} 4 & 1 & -1 & | & 4 \\ 8 & 2 & -2 & | & 8 \end{bmatrix} \xrightarrow{\frac{1}{4}R_1} \begin{bmatrix} 1 & \frac{1}{4} & \frac{1}{4} & | & 1 \\ 8 & 2 & -2 & | & 8 \end{bmatrix} \xrightarrow{R_2-8R_1} \begin{bmatrix} 1 & \frac{1}{4} & -\frac{1}{4} & | & 1 \\ 0 & 0 & 0 & | & 0 \end{bmatrix}$

We conclude that the infinitely many solutions are given by $(1-\frac{1}{4}s+\frac{1}{4}t, s, t)$, where s and t are parameters.

29. $\begin{bmatrix} 2 & 1 & -3 & | & 1 \\ 1 & -1 & 2 & | & 1 \\ 5 & -2 & 3 & | & 6 \end{bmatrix} \xrightarrow{R_1 \leftrightarrow R_2} \begin{bmatrix} 1 & -1 & 2 & | & 1 \\ 2 & 1 & -3 & | & 1 \\ 5 & -2 & 3 & | & 6 \end{bmatrix} \xrightarrow[R_3-5R_1]{R_2-2R_1} \begin{bmatrix} 1 & -1 & 2 & | & 1 \\ 0 & 3 & -7 & | & -1 \\ 0 & 3 & -7 & | & 1 \end{bmatrix} \xrightarrow{\frac{1}{3}R_2}$

$\begin{bmatrix} 1 & -1 & 2 & | & 1 \\ 0 & 1 & -\frac{7}{3} & | & -\frac{1}{3} \\ 0 & 3 & -7 & | & 1 \end{bmatrix} \xrightarrow[R_3-3R_2]{R_1+R_2} \begin{bmatrix} 1 & 0 & -\frac{1}{3} & | & \frac{2}{3} \\ 0 & 1 & -\frac{7}{3} & | & -\frac{1}{3} \\ 0 & 0 & 0 & | & 2 \end{bmatrix}.$

This last row implies that $0 = 2$, which is impossible. We conclude that the system of equations is inconsistent and has no solution.

31. $\begin{bmatrix} 1 & 2 & -1 & | & -4 \\ 2 & 1 & 1 & | & 7 \\ 1 & 3 & 2 & | & 7 \\ 1 & -3 & 1 & | & 9 \end{bmatrix} \xrightarrow[R_4-R_1]{\substack{R_2-2R_1 \\ R_3-R_1}} \begin{bmatrix} 1 & 2 & -1 & | & -4 \\ 0 & -3 & 3 & | & 15 \\ 0 & 1 & 3 & | & 11 \\ 0 & -5 & 2 & | & 13 \end{bmatrix} \xrightarrow{-\frac{1}{3}R_2} \begin{bmatrix} 1 & 2 & -1 & | & -4 \\ 0 & 1 & -1 & | & -5 \\ 0 & 1 & 3 & | & 11 \\ 0 & -5 & 2 & | & 13 \end{bmatrix}$

$\xrightarrow[R_4+5R_2]{\substack{R_1-2R_2 \\ R_3-R_2}} \begin{bmatrix} 1 & 0 & 1 & | & 6 \\ 0 & 1 & -1 & | & -5 \\ 0 & 0 & 4 & | & 16 \\ 0 & 0 & -3 & | & -12 \end{bmatrix} \xrightarrow{\frac{1}{4}R_3} \begin{bmatrix} 1 & 0 & 1 & | & 6 \\ 0 & 1 & -1 & | & -5 \\ 0 & 0 & 1 & | & 4 \\ 0 & 0 & -3 & | & -12 \end{bmatrix} \xrightarrow[R_4+3R_3]{\substack{R_1-R_3 \\ R_2+R_3}} \begin{bmatrix} 1 & 0 & 0 & | & 2 \\ 0 & 1 & 0 & | & -1 \\ 0 & 0 & 1 & | & 4 \\ 0 & 0 & 0 & | & 0 \end{bmatrix}.$

We conclude that the solution of the system is $(2,-1,4)$.

33. Let x, y, and z represent the number of compact, mid-sized, and full-size cars,

respectively, to be purchased. Then the problem can be solved by solving the system

$$x + y + z = 60$$
$$10000x + 16000y + 22000z = 840000 \ .$$

Using the Gauss-Jordan method, we have

$$\begin{bmatrix} 1 & 1 & 1 & 60 \\ 10000 & 16000 & 22000 & 840000 \end{bmatrix} \xrightarrow{R_2 - 10,000R_1} \begin{bmatrix} 1 & 1 & 1 & 60 \\ 0 & 6000 & 12000 & 240000 \end{bmatrix}$$

$$\xrightarrow{\frac{1}{6000}R_2} \begin{bmatrix} 1 & 1 & 1 & 60 \\ 0 & 1 & 2 & 40 \end{bmatrix} \xrightarrow{R_1 - R_2} \begin{bmatrix} 1 & 0 & -1 & 20 \\ 0 & 1 & 2 & 40 \end{bmatrix}$$

and we conclude that the solution is $(20 + z, 40 - 2z, z)$. Letting $z = 5$, we see that one possible solution is $(25,30,5)$; that is Hartman should buy 25 compact, 30 mid-sized cars, and 5 full-sized cars. Letting $z = 10$, we see that another possible solution is $(30,20,10)$; that is, 30 compact cars, 20 mid-sized cars, and 10 full-sized cars.

35. Let $x, y,$ and z denote the number of ounces of Food I, Food II, and Food III, respectively, that the dietician includes in the meal. Then the problem can be solved by solving the system

$$400x + 1200y + 800z = 8800$$
$$110x + 570y + 340z = 2160$$
$$90x + 30y + 60z = 1020.$$

Using the Gauss-Jordan method, we have

$$\begin{bmatrix} 400 & 1200 & 800 & 8800 \\ 110 & 570 & 340 & 2160 \\ 90 & 30 & 60 & 1020 \end{bmatrix} \xrightarrow{\frac{1}{400}R_1} \begin{bmatrix} 1 & 3 & 2 & 22 \\ 110 & 570 & 340 & 2160 \\ 90 & 30 & 60 & 1020 \end{bmatrix} \xrightarrow[R_3 - 90R_1]{R_2 - 110R_1}$$

$$\begin{bmatrix} 1 & 3 & 2 & 22 \\ 0 & 240 & 120 & -260 \\ 0 & -240 & -120 & -960 \end{bmatrix} \xrightarrow{\frac{1}{240}R_2} \begin{bmatrix} 1 & 3 & 2 & 22 \\ 0 & 1 & \frac{1}{2} & -\frac{13}{12} \\ 0 & -240 & -120 & -960 \end{bmatrix} \xrightarrow[R_3 + 240R_2]{R_1 - 3R_2}$$

$$\begin{bmatrix} 1 & 0 & \frac{1}{2} & \frac{101}{4} \\ 0 & 1 & \frac{1}{2} & -\frac{13}{12} \\ 0 & 0 & 0 & -1220 \end{bmatrix}.$$

This last row implies that $0 = -1220$, which is impossible. We conclude that the system of equations is inconsistent and has no solution--that is, the dietician cannot prepare a meal from these foods and meet the given requirements.

37. a.

$$\begin{aligned} x_1 - x_2 &= 200 \\ x_1 \qquad\qquad -x_5 &= 100 \\[6pt] -x_2 + x_3 \qquad\quad + x_6 &= 600 \\ -x_3 + x_4 &= 200 \\ x_4 - x_5 + x_6 &= 700. \end{aligned}$$

b.

$$\begin{bmatrix} 1 & -1 & 0 & 0 & 0 & 0 & 200 \\ 1 & 0 & 0 & 0 & -1 & 0 & 100 \\ 0 & -1 & 1 & 0 & 0 & 1 & 600 \\ 0 & 0 & -1 & 1 & 0 & 0 & 200 \\ 0 & 0 & 0 & 1 & -1 & 1 & 700 \end{bmatrix} \xrightarrow{R_2 - R_1} \begin{bmatrix} 1 & -1 & 0 & 0 & 0 & 0 & 200 \\ 0 & 1 & 0 & 0 & -1 & 0 & -100 \\ 0 & -1 & 1 & 0 & 0 & 1 & 600 \\ 0 & 0 & -1 & 1 & 0 & 0 & 200 \\ 0 & 0 & 0 & 1 & -1 & 1 & 700 \end{bmatrix} \xrightarrow[R_1 + R_2]{R_3 + R_2}$$

$$\begin{bmatrix} 1 & 0 & 0 & 0 & -1 & 0 & 100 \\ 0 & 1 & 0 & 0 & -1 & 0 & -100 \\ 0 & 0 & 1 & 0 & -1 & 1 & 500 \\ 0 & 0 & -1 & 1 & 0 & 0 & 200 \\ 0 & 0 & 0 & 1 & -1 & 1 & 700 \end{bmatrix} \xrightarrow{R_4 + R_3} \begin{bmatrix} 1 & 0 & 0 & 0 & -1 & 0 & 100 \\ 0 & 1 & 0 & 0 & -1 & 0 & -100 \\ 0 & 0 & 1 & 0 & -1 & 1 & 500 \\ 0 & 0 & 0 & 1 & -1 & 1 & 700 \\ 0 & 0 & 0 & 1 & -1 & 1 & 700 \end{bmatrix} \xrightarrow{R_5 - R_4}$$

$$\begin{bmatrix} 1 & 0 & 0 & 0 & -1 & 0 & 100 \\ 0 & 1 & 0 & 0 & 1 & 0 & -100 \\ 0 & 0 & 1 & 0 & -1 & 1 & 500 \\ 0 & 0 & 0 & 1 & -1 & 1 & 700 \\ 0 & 0 & 0 & 0 & 0 & 0 & 0 \end{bmatrix}.$$

We conclude that the solution is
$$(s+100,\ s-100,\ s-t+500,\ s-t+700,\ s,\ t).$$
Taking $s = 150$ and $t = 50$, we see that one possible traffic pattern is
$$(250, 50, 600, 800, 150, 50).$$
Similarly, taking $s = 200$, and $t = 100$, we see that another possible traffic pattern is
$$(300, 100, 600, 800, 200, 100).$$
c. Taking $t = 0$ and $s = 200$, we see that another possible traffic pattern is

(300, 100, 700, 900, 200, 0).

39. We solve the given system by using the Gauss-Jordan method. We have

$$\begin{bmatrix} 2 & 3 & | & 2 \\ 1 & 4 & | & 6 \\ 5 & k & | & 2 \end{bmatrix} \xrightarrow{R_1 \leftrightarrow R_2} \begin{bmatrix} 1 & 4 & | & 6 \\ 2 & 3 & | & 2 \\ 5 & k & | & 2 \end{bmatrix} \xrightarrow[R_3-5R_1]{R_2-2R_1} \begin{bmatrix} 1 & 4 & | & 6 \\ 0 & -5 & | & -10 \\ 0 & k-20 & | & -28 \end{bmatrix} \xrightarrow{-\frac{1}{5}R_2}$$

$$\begin{bmatrix} 1 & 4 & | & 6 \\ 0 & 1 & | & 2 \\ 0 & k-20 & | & -28 \end{bmatrix} \xrightarrow[R_3+aR_2]{R_1-4R_2} \begin{bmatrix} 1 & 0 & | & -2 \\ 0 & 1 & | & 2 \\ 0 & k+a-20 & | & -28+2a \end{bmatrix}$$

From the last matrix, we see that the system has a solution if and only if $x = -2$, $y = 2$, and

$$-28 + 2a = 0, \text{ or } a = 14$$

and $k + a - 20 = k - 6 = 0$, or $k = 6$.

(All the entries in the last row of the matrix must be equal to zero.)

41. False. Such a system cannot have a unique solution.

USING TECHNOLOGY EXERCISES 2.3, page 109

1. $(1+t, 2+t, t)$; t, a parameter   3. $\left(-\frac{17}{7} + \frac{6}{7}t, 3-t, -\frac{18}{7} + \frac{1}{7}t, t\right)$ 5.   No solution

EXERCISES 2.4, page 117

1. The size of A is 4×4; the size of B is 4×3; the size of C is 1×5, and the size of D is 4×1.

3. These are entries of the matrix B. The entry b_{13} refers to the entry in the first row and third column and is equal to 2. Similarly, $b_{31} = 3$, and $b_{43} = 8$.

5. The column matrix is the matrix D. The transpose of the matrix D is
$$D^T = [1 \ 3 \ -2 \ 0].$$

7. A is of size 3×2; B is of size 3×2; C and D are of size 3×3.

9.
$$A + B = \begin{bmatrix} -1 & 2 \\ 3 & -2 \\ 4 & 0 \end{bmatrix} + \begin{bmatrix} 2 & 4 \\ 3 & 1 \\ -2 & 2 \end{bmatrix} = \begin{bmatrix} 1 & 6 \\ 6 & -1 \\ 2 & 2 \end{bmatrix}.$$

11.
$$\begin{bmatrix} 3 & -1 & 0 \\ 2 & -2 & 3 \\ 4 & 6 & 2 \end{bmatrix} - \begin{bmatrix} 2 & -2 & 4 \\ 3 & 6 & 2 \\ -2 & 3 & 1 \end{bmatrix} = \begin{bmatrix} 1 & 1 & -4 \\ -1 & -8 & 1 \\ 6 & 3 & 1 \end{bmatrix}.$$

13.
$$\begin{bmatrix} 6 & 3 & 8 \\ 4 & 5 & 6 \end{bmatrix} - \begin{bmatrix} 3 & -2 & -1 \\ 0 & -5 & -7 \end{bmatrix} = \begin{bmatrix} 3 & 5 & 9 \\ 4 & 10 & 13 \end{bmatrix}.$$

15.
$$\begin{bmatrix} 1 & 4 & -5 \\ 3 & -8 & 6 \end{bmatrix} + \begin{bmatrix} 4 & 0 & -2 \\ 3 & 6 & 5 \end{bmatrix} - \begin{bmatrix} 2 & 8 & 9 \\ -11 & 2 & -5 \end{bmatrix} = \begin{bmatrix} 3 & -4 & -16 \\ 17 & -4 & 16 \end{bmatrix}.$$

17.
$$\begin{bmatrix} 1.2 & 4.5 & -4.2 \\ 8.2 & 6.3 & -3.2 \end{bmatrix} - \begin{bmatrix} 3.1 & 1.5 & -3.6 \\ 2.2 & -3.3 & -4.4 \end{bmatrix} = \begin{bmatrix} -1.9 & 3.0 & -0.6 \\ 6.0 & 9.6 & 1.2 \end{bmatrix}.$$

19.
$$\frac{1}{2}\begin{bmatrix} 1 & 0 & 0 & -4 \\ 3 & 0 & -1 & 6 \\ -2 & 1 & -4 & 2 \end{bmatrix} + \frac{4}{3}\begin{bmatrix} 3 & 0 & -1 & 4 \\ -2 & 1 & -6 & 2 \\ 8 & 2 & 0 & -2 \end{bmatrix} - \frac{1}{3}\begin{bmatrix} 3 & -9 & -1 & 0 \\ 6 & 2 & 0 & -6 \\ 0 & 1 & -3 & 1 \end{bmatrix}$$

$$= \begin{bmatrix} \frac{7}{2} & 3 & -1 & \frac{10}{3} \\ -\frac{19}{6} & \frac{2}{3} & -\frac{17}{2} & \frac{23}{3} \\ \frac{29}{3} & \frac{17}{6} & -1 & -2 \end{bmatrix}.$$

21.
$$\begin{bmatrix} 2x-2 & 3 & 2 \\ 2 & 4 & y-2 \\ 2z & -3 & 2 \end{bmatrix} = \begin{bmatrix} 3 & u & 2 \\ 2 & 4 & 5 \\ 4 & -3 & 2 \end{bmatrix}.$$

Now, by the definition of equality of matrices,

$$u = 3$$

$$2x - 2 = 3 \text{ and } 2x = 5, \text{ or } x = 5/2,$$

$$y - 2 = 5, \text{ and } y = 7,$$

$$2z = 4, \text{ and } z = 2.$$

23. $\begin{bmatrix} 1 & x \\ 2y & -3 \end{bmatrix} - 4\begin{bmatrix} 2 & -2 \\ 0 & 3 \end{bmatrix} = \begin{bmatrix} 3z & 10 \\ 4 & -u \end{bmatrix}; \begin{bmatrix} -7 & x+8 \\ 2y & -15 \end{bmatrix} = \begin{bmatrix} 3z & 10 \\ 4 & -u \end{bmatrix}.$

Now, by the definition of equality of matrices,

$$-u = -15, \text{ so } u = 15$$

$$x + 8 = 10, \text{ so } x = 2$$

$$2y = 4, \text{ so } y = 2$$

$$3z = -7, \text{ so } z = -7/3.$$

25. To verify the Commutative Law for matrix addition, let us show that $A + B = B + A$.

Now, $A + B = \begin{bmatrix} 2 & -4 & 3 \\ 4 & 2 & 1 \end{bmatrix} + \begin{bmatrix} 4 & -3 & 2 \\ 1 & 0 & 4 \end{bmatrix} = \begin{bmatrix} 6 & -7 & 5 \\ 5 & 2 & 5 \end{bmatrix}$

$$= \begin{bmatrix} 4 & -3 & 2 \\ 1 & 0 & 4 \end{bmatrix} + \begin{bmatrix} 2 & -4 & 3 \\ 4 & 2 & 1 \end{bmatrix} = B + A.$$

27. $(3+5)A = 8A = 8\begin{bmatrix} 3 & 1 \\ 2 & 4 \\ -4 & 0 \end{bmatrix} = \begin{bmatrix} 24 & 8 \\ 16 & 32 \\ -32 & 0 \end{bmatrix} = 3\begin{bmatrix} 3 & 1 \\ 2 & 4 \\ -4 & 0 \end{bmatrix} + 5\begin{bmatrix} 3 & 1 \\ 2 & 4 \\ -4 & 0 \end{bmatrix}$

$$= 3A + 5A.$$

29. $4(A + B) = 4\left(\begin{bmatrix} 3 & 1 \\ 2 & 4 \\ -4 & 0 \end{bmatrix} + \begin{bmatrix} 1 & 2 \\ -1 & 0 \\ 3 & 2 \end{bmatrix} \right) = 4\begin{bmatrix} 4 & 3 \\ 1 & 4 \\ -1 & 2 \end{bmatrix} = \begin{bmatrix} 16 & 12 \\ 4 & 16 \\ -4 & 8 \end{bmatrix}$

$$4A + 4B = 4\begin{bmatrix} 3 & 1 \\ 2 & 4 \\ -4 & 0 \end{bmatrix} + 4\begin{bmatrix} 1 & 2 \\ -1 & 0 \\ 3 & 2 \end{bmatrix} = \begin{bmatrix} 16 & 12 \\ 4 & 16 \\ -4 & 8 \end{bmatrix}.$$

31. $[3 \quad 2 \quad -1 \quad 5]^T = \begin{bmatrix} 3 \\ 2 \\ -1 \\ 5 \end{bmatrix}.$

33. $\begin{bmatrix} 1 & -1 & 2 \\ 3 & 4 & 2 \\ 0 & 1 & 0 \end{bmatrix}^T = \begin{bmatrix} 1 & 3 & 0 \\ -1 & 4 & 1 \\ 2 & 2 & 0 \end{bmatrix}.$

35.

	1	2	3	4
Mr. Cross	220	215	210	205
Mr. Jones	220	210	200	195
Mr. Smith	215	205	195	190

37.　a.　$D = A + B - C$

$$= \begin{bmatrix} 2820 & 1470 & 1120 \\ 1030 & 520 & 480 \\ 1170 & 540 & 460 \end{bmatrix} + \begin{bmatrix} 260 & 120 & 110 \\ 140 & 60 & 50 \\ 120 & 70 & 50 \end{bmatrix} - \begin{bmatrix} 120 & 80 & 80 \\ 70 & 30 & 40 \\ 60 & 20 & 40 \end{bmatrix}$$

$$= \begin{bmatrix} 2960 & 1510 & 1150 \\ 1100 & 550 & 490 \\ 1230 & 590 & 470 \end{bmatrix}.$$

　　b.　$E = 1.1D = 1.1 \begin{bmatrix} 2960 & 1510 & 1150 \\ 1100 & 550 & 490 \\ 1230 & 590 & 470 \end{bmatrix} = \begin{bmatrix} 3256 & 1661 & 1265 \\ 1210 & 605 & 539 \\ 1353 & 649 & 517 \end{bmatrix}.$

39. True. Each element in $A + B$ is obtained by adding together the corresponding elements in A and B. Therefore, the matrix $c(A+B)$ is obtained by multiplying each element in $A + B$ by c. On the other hand, cA is obtained by multiplying each element in A by c and cB is obtained by multiplying each element in B by c and $cA + cB$ is obtained by adding the corresponding elements in cA and cB. Thus $c(A+B) = cA + cB$.

41. False. Take $\begin{bmatrix} 1 & 2 \\ 3 & 4 \end{bmatrix}$ and $c = 2$. Then

$$cA = 2\begin{bmatrix} 1 & 2 \\ 3 & 4 \end{bmatrix} = \begin{bmatrix} 2 & 4 \\ 6 & 8 \end{bmatrix} \text{ and } (cA)^T = \begin{bmatrix} 2 & 6 \\ 4 & 8 \end{bmatrix}.$$

On the other hand, $\dfrac{1}{c}A^T = \dfrac{1}{2}\begin{bmatrix} 1 & 3 \\ 2 & 4 \end{bmatrix} = \begin{bmatrix} \frac{1}{2} & \frac{3}{2} \\ 1 & 2 \end{bmatrix} \neq (cA)^T$

USING TECHNOLOGY EXERCISES 2.4, page 121

1. $\begin{bmatrix} 15 & 38.75 & -67.5 & 33.75 \\ 51.25 & 40 & 52.5 & -38.75 \\ 21.25 & 35 & -65 & 105 \end{bmatrix}$

3. $\begin{bmatrix} -5 & 6.3 & -6.8 & 3.9 \\ 1 & 0.5 & 5.4 & -4.8 \\ 0.5 & 4.2 & -3.5 & 5.6 \end{bmatrix}$

5. $\begin{bmatrix} 16.44 & -3.65 & -3.66 & 0.63 \\ 12.77 & 10.64 & 2.58 & 0.05 \\ 5.09 & 0.28 & -10.84 & 17.64 \end{bmatrix}$

7. $\begin{bmatrix} 22.2 & -0.3 & -12 & 4.5 \\ 21.6 & 17.7 & 9 & -4.2 \\ 8.7 & 4.2 & -20.7 & 33.6 \end{bmatrix}$

EXERCISES 2.5, page 132

1. $(2 \times 3)(3 \times 5)$ so AB has order 2×5.

$(3 \times 5)(2 \times 3)$ so BA is not defined.

3. $(1 \times 7)\,(7 \times 1)$ so AB has order 1×1.

$(7 \times 1)\,(1 \times 7)$ so BA has order 7×7.

5. If AB and BA are defined then $n = s$ and $m = t$.

7. $\begin{bmatrix} 1 & 2 \\ 3 & 0 \end{bmatrix} \begin{bmatrix} 1 \\ -1 \end{bmatrix} = \begin{bmatrix} -1 \\ 3 \end{bmatrix}$

9. $\begin{bmatrix} 3 & 1 & 2 \\ -1 & 2 & 4 \end{bmatrix} \begin{bmatrix} 4 \\ 1 \\ -2 \end{bmatrix} = \begin{bmatrix} 9 \\ -10 \end{bmatrix}$

11. $\begin{bmatrix} -1 & 2 \\ 3 & 1 \end{bmatrix} \begin{bmatrix} 2 & 4 \\ 3 & 1 \end{bmatrix} = \begin{bmatrix} 4 & -2 \\ 9 & 13 \end{bmatrix}$

13. $\begin{bmatrix} 2 & 1 & 2 \\ 3 & 2 & 4 \end{bmatrix} \begin{bmatrix} -1 & 2 \\ 4 & 3 \\ 0 & 1 \end{bmatrix} = \begin{bmatrix} 2 & 9 \\ 5 & 16 \end{bmatrix}$

15. $\begin{bmatrix} 0.1 & 0.9 \\ 0.2 & 0.8 \end{bmatrix} \begin{bmatrix} 1.2 & 0.4 \\ 0.5 & 2.1 \end{bmatrix} = \begin{bmatrix} 0.1(1.2)+0.9(0.5) & 0.1(0.4)+0.9(2.1) \\ 0.2(1.2)+0.8(0.5) & 0.2(0.4)+0.8(2.1) \end{bmatrix} = \begin{bmatrix} 0.57 & 1.93 \\ 0.64 & 1.76 \end{bmatrix}$

17. $\begin{bmatrix} 6 & -3 & 0 \\ -2 & 1 & -8 \\ 4 & -4 & 9 \end{bmatrix} \begin{bmatrix} 1 & 0 & 0 \\ 0 & 1 & 0 \\ 0 & 0 & 1 \end{bmatrix} = \begin{bmatrix} 6 & -3 & 0 \\ -2 & 1 & -8 \\ 4 & -4 & 9 \end{bmatrix}.$

19. $\begin{bmatrix} 3 & 0 & -2 & 1 \\ 1 & 2 & 0 & -1 \end{bmatrix} \begin{bmatrix} 2 & 1 & -1 \\ -1 & 2 & 0 \\ 0 & 0 & 1 \\ -1 & -2 & 2 \end{bmatrix} = \begin{bmatrix} 5 & 1 & -3 \\ 1 & 7 & -3 \end{bmatrix}.$

21. $4\begin{bmatrix} 1 & -2 & 0 \\ 2 & -1 & 1 \\ 3 & 0 & -1 \end{bmatrix} \begin{bmatrix} 1 & 3 & 1 \\ 1 & 4 & 0 \\ 0 & 1 & -2 \end{bmatrix} = \begin{bmatrix} -4 & -20 & 4 \\ 4 & 12 & 0 \\ 12 & 32 & 20 \end{bmatrix}$

23. $\begin{bmatrix} 1 & 0 \\ 0 & 1 \end{bmatrix} \begin{bmatrix} 4 & -3 & 2 \\ 7 & 1 & -5 \end{bmatrix} \begin{bmatrix} 1 & 0 & 0 \\ 0 & 1 & 0 \\ 0 & 0 & 1 \end{bmatrix} = \begin{bmatrix} 1 & 0 \\ 0 & 1 \end{bmatrix} \begin{bmatrix} 4 & -3 & 2 \\ 7 & 1 & -5 \end{bmatrix} = \begin{bmatrix} 4 & -3 & 2 \\ 7 & 1 & -5 \end{bmatrix}.$

25. To verify the associative law for matrix multiplication, we will show that $(AB)C = A(BC)$.

$$AB = \begin{bmatrix} 1 & 0 & -2 \\ 1 & -3 & 2 \\ -2 & 1 & 1 \end{bmatrix} \begin{bmatrix} 3 & 1 & 0 \\ 2 & 2 & 0 \\ 1 & -3 & -1 \end{bmatrix} = \begin{bmatrix} 1 & 7 & 2 \\ -1 & -11 & -2 \\ -3 & -3 & -1 \end{bmatrix}$$

$$(AB)C = \begin{bmatrix} 1 & 7 & 2 \\ -1 & -11 & -2 \\ -3 & -3 & -1 \end{bmatrix} \begin{bmatrix} 2 & -1 & 0 \\ 1 & -1 & 2 \\ 3 & -2 & 1 \end{bmatrix} = \begin{bmatrix} 15 & -12 & 16 \\ -19 & 16 & -24 \\ -12 & 8 & -7 \end{bmatrix}$$

$$BC = \begin{bmatrix} 3 & 1 & 0 \\ 2 & 2 & 0 \\ 1 & -3 & -1 \end{bmatrix} \begin{bmatrix} 2 & -1 & 0 \\ 1 & -1 & 2 \\ 3 & -2 & 1 \end{bmatrix} = \begin{bmatrix} 7 & -4 & 2 \\ 6 & -4 & 4 \\ -4 & 4 & -7 \end{bmatrix}$$

$$A(BC) = \begin{bmatrix} 1 & 0 & -2 \\ 1 & -3 & 2 \\ -2 & 1 & 1 \end{bmatrix} \begin{bmatrix} 7 & -4 & 2 \\ 6 & -4 & 4 \\ -4 & 4 & -7 \end{bmatrix} = \begin{bmatrix} 15 & -12 & 16 \\ -19 & 16 & -24 \\ -12 & 8 & -7 \end{bmatrix}.$$

27.
$$AB = \begin{bmatrix} 1 & 2 \\ 3 & 4 \end{bmatrix} \begin{bmatrix} 2 & 1 \\ 4 & 3 \end{bmatrix} = \begin{bmatrix} 10 & 7 \\ 22 & 15 \end{bmatrix}$$

$$BA = \begin{bmatrix} 2 & 1 \\ 4 & 3 \end{bmatrix} \begin{bmatrix} 1 & 2 \\ 3 & 4 \end{bmatrix} = \begin{bmatrix} 5 & 8 \\ 13 & 20 \end{bmatrix}$$

Therefore, $AB \neq BA$ and matrix multiplication is not commutative.

29.
$$AB = \begin{bmatrix} 3 & 0 \\ 8 & 0 \end{bmatrix} \begin{bmatrix} 0 & 0 \\ 4 & 5 \end{bmatrix} = \begin{bmatrix} 0 & 0 \\ 0 & 0 \end{bmatrix}$$

$AB = 0$, but neither A nor B is the zero matrix. Therefore, $AB = 0$, does not imply that A or B is the zero matrix.

31.
$$\begin{bmatrix} a & b \\ c & d \end{bmatrix} \begin{bmatrix} 1 & 0 \\ -1 & 3 \end{bmatrix} = \begin{bmatrix} a-b & 3b \\ c-d & 3d \end{bmatrix} = \begin{bmatrix} -1 & -3 \\ 3 & 6 \end{bmatrix}$$

Then $3b = -3,$ and $b = -1$
         $3d = 6,$ and $d = 2$
         $a - b = -1,$ and $a = b - 1 = -2.$
         $c - d = 3,$ and $c = d + 3 = 5$

Therefore, $A = \begin{bmatrix} -2 & -1 \\ 5 & 2 \end{bmatrix}.$

33. a.

$$A^T = \begin{bmatrix} 2 & 5 \\ 4 & -6 \end{bmatrix} \text{ and } (A^T)^T = \begin{bmatrix} 2 & 4 \\ 5 & -6 \end{bmatrix} = A$$

b. $(A+B)^T = \begin{bmatrix} 6 & 12 \\ -2 & -3 \end{bmatrix}^T = \begin{bmatrix} 6 & -2 \\ 12 & -3 \end{bmatrix}$

$$A^T + B^T = \begin{bmatrix} 2 & 5 \\ 4 & -6 \end{bmatrix} + \begin{bmatrix} 4 & -7 \\ 8 & 3 \end{bmatrix} = \begin{bmatrix} 6 & -2 \\ 12 & -3 \end{bmatrix}$$

c. $AB = \begin{bmatrix} 2 & 4 \\ 5 & -6 \end{bmatrix}\begin{bmatrix} 4 & 8 \\ -7 & 3 \end{bmatrix} = \begin{bmatrix} -20 & 28 \\ 62 & 22 \end{bmatrix},$ so $(AB)^T = \begin{bmatrix} -20 & 62 \\ 28 & 22 \end{bmatrix}.$

$$B^T A^T = \begin{bmatrix} 4 & -7 \\ 8 & 3 \end{bmatrix}\begin{bmatrix} 2 & 5 \\ 4 & -6 \end{bmatrix} = \begin{bmatrix} -20 & 62 \\ 28 & 22 \end{bmatrix} = (AB)^T$$

35. The given system of linear equations can be represented by the matrix equation
$AX = B$, where

$$A = \begin{bmatrix} 2 & -3 \\ 3 & -4 \end{bmatrix}, \quad X = \begin{bmatrix} x \\ y \end{bmatrix}, \quad \text{and } B = \begin{bmatrix} 7 \\ 8 \end{bmatrix}.$$

37. The given system of linear equations can be represented by the matrix equation
$AX = B$, where

$$A = \begin{bmatrix} 2 & -3 & 4 \\ 0 & 2 & -3 \\ 1 & -1 & 2 \end{bmatrix}, \quad X = \begin{bmatrix} x \\ y \\ z \end{bmatrix}, \quad B = \begin{bmatrix} 6 \\ 7 \\ 4 \end{bmatrix}.$$

39. The given system of linear equations can be represented by the matrix equation $AX = B$, where

$$A = \begin{bmatrix} -1 & 1 & 1 \\ 2 & -1 & -1 \\ -3 & 2 & 4 \end{bmatrix}, \quad X = \begin{bmatrix} x_1 \\ x_2 \\ x_3 \end{bmatrix}, \quad B = \begin{bmatrix} 0 \\ 2 \\ 4 \end{bmatrix}.$$

41. a.
$$AB = \begin{bmatrix} 200 & 300 & 100 & 200 \\ 100 & 200 & 400 & 0 \end{bmatrix} \begin{bmatrix} 54 \\ 48 \\ 98 \\ 82 \end{bmatrix} = \begin{bmatrix} 51,400 \\ 54,200 \end{bmatrix}$$

b. The first entry shows that William's total stock holdings are $51,400, while the second entry shows that Michael's stockholdings are $54,200.

43. The column vector that represents the profit for each type of house is

$$B = \begin{bmatrix} 20,000 \\ 22,000 \\ 25,000 \\ 30,000 \end{bmatrix}.$$

The column vector that gives the total profit for Bond Brothers is

$$AB = \begin{bmatrix} 60 & 80 & 120 & 40 \\ 20 & 30 & 60 & 10 \\ 10 & 15 & 30 & 5 \end{bmatrix} \begin{bmatrix} 20,000 \\ 22,000 \\ 25,000 \\ 30,000 \end{bmatrix}$$

$$= \begin{bmatrix} 7,160,000 \\ 2,860,000 \\ 1,430,000 \end{bmatrix}.$$

Therefore, Bond Brothers expects to make $7,160,000 in New York, $2,860,000 in Connecticut, and $1,430,000 in Massachusetts, and the total profit is $11,450,000.

45.
$$BA = \begin{bmatrix} 30,000 & 40,000 & 20,000 \end{bmatrix} \begin{matrix} D & R & I \\ \begin{bmatrix} 0.50 & 0.30 & 0.20 \\ 0.45 & 0.40 & 0.15 \\ 0.40 & 0.50 & 0.10 \end{bmatrix} \end{matrix}$$

$$\begin{array}{ccc} D & R & I \end{array}$$
$$= \begin{bmatrix} 41{,}000 & 35{,}000 & 14{,}000 \end{bmatrix}$$

47. $AB = \begin{bmatrix} 2700 & 3000 \\ 800 & 700 \\ 500 & 300 \end{bmatrix} \begin{bmatrix} 0.25 & 0.20 & 0.30 & 0.25 \\ 0.30 & 0.35 & 0.25 & 0.10 \end{bmatrix} = \begin{bmatrix} 1575 & 1590 & 1560 & 975 \\ 410 & 405 & 415 & 270 \\ 215 & 205 & 225 & 155 \end{bmatrix}$

49. $AC = \begin{bmatrix} 80 & 60 & 40 \end{bmatrix} \begin{bmatrix} 0.34 \\ 0.42 \\ 0.48 \end{bmatrix} = 71.6 \qquad BD = \begin{bmatrix} 300 & 150 & 250 \end{bmatrix} \begin{bmatrix} 0.24 \\ 0.31 \\ 0.35 \end{bmatrix} = 206$

$AC + BD = \begin{bmatrix} 277.60 \end{bmatrix}$, or \$277.60. It represents Cindy's long distance bill for phone calls to those 3 cities.

51. a. $MA^T = \begin{bmatrix} 400 & 1200 & 800 \\ 110 & 570 & 340 \\ 90 & 30 & 60 \end{bmatrix} \begin{bmatrix} 7 \\ 1 \\ 6 \end{bmatrix} = \begin{bmatrix} 8800 \\ 3380 \\ 1020 \end{bmatrix}$.

The amounts of vitamin A, vitamin C, and calcium taken by a girl in the first meal are 8800, 3380, and 1020 units respectively.

b. $MB^T = \begin{bmatrix} 400 & 1200 & 800 \\ 110 & 570 & 340 \\ 90 & 30 & 60 \end{bmatrix} \begin{bmatrix} 9 \\ 3 \\ 2 \end{bmatrix} = \begin{bmatrix} 8800 \\ 3380 \\ 1020 \end{bmatrix}$

The amounts of vitamin A, vitamin C, and calcium taken by a girl in the second meal are 8,800, 3380, and 1020 units, respectively.

c. $M(A+B)^T = \begin{bmatrix} 400 & 1200 & 800 \\ 110 & 570 & 340 \\ 90 & 30 & 60 \end{bmatrix} \begin{bmatrix} 16 \\ 4 \\ 8 \end{bmatrix} = \begin{bmatrix} 17{,}600 \\ 6{,}760 \\ 2{,}040 \end{bmatrix}$.

The amounts of vitamin A, vitamin C, and calcium taken by a girl in the two meals are 17,600, 6,760, and 2,040 units respectively.

53. False. Let A be a matrix of order 2×3 and let B be a matrix of order 3×2. Then AB and BA are both defined. But, evidently, neither A nor B is a square matrix.

USING TECHNOLOGY EXERCISES 2.5, page 136

1. $\begin{bmatrix} 18.66 & 15.2 & -12 \\ 24.48 & 41.88 & 89.82 \\ 15.39 & 7.16 & -1.25 \end{bmatrix}$

3. $\begin{bmatrix} 20.09 & 20.61 & -1.3 \\ 44.42 & 71.6 & 64.89 \\ 20.97 & 7.17 & -60.65 \end{bmatrix}$

5. $\begin{bmatrix} 32.89 & 13.63 & -57.17 \\ -12.85 & -8.37 & 256.92 \\ 13.48 & 14.29 & 181.64 \end{bmatrix}$

7. $\begin{bmatrix} 128.59 & 123.08 & -32.50 \\ 246.73 & 403.12 & 481.52 \\ 125.06 & 47.01 & -264.81 \end{bmatrix}$

9. $\begin{bmatrix} 87 & 68 & 110 & 82 \\ 119 & 176 & 221 & 143 \\ 51 & 128 & 142 & 94 \\ 28 & 174 & 174 & 112 \end{bmatrix}$
$\begin{bmatrix} 113 & 117 & 72 & 101 & 90 \\ 72 & 85 & 36 & 72 & 76 \\ 81 & 69 & 76 & 87 & 30 \\ 133 & 157 & 56 & 121 & 146 \\ 154 & 157 & 94 & 127 & 122 \end{bmatrix}$

11. $\begin{bmatrix} 170 & 18.1 & 133.1 & -106.3 & 341.3 \\ 349 & 226.5 & 324.1 & 164 & 506.4 \\ 245.2 & 157.7 & 231.5 & 125.5 & 312.9 \\ 310 & 245.2 & 291 & 274.3 & 354.2 \end{bmatrix}$

EXERCISES 2.6, page 149

1. $\begin{bmatrix} 1 & -3 \\ 1 & -2 \end{bmatrix}\begin{bmatrix} -2 & 3 \\ -1 & 1 \end{bmatrix} = \begin{bmatrix} 1 & 0 \\ 0 & 1 \end{bmatrix}; \quad \begin{bmatrix} -2 & 3 \\ -1 & 1 \end{bmatrix}\begin{bmatrix} 1 & -3 \\ 1 & -2 \end{bmatrix} = \begin{bmatrix} 1 & 0 \\ 0 & 1 \end{bmatrix}$

3. $\begin{bmatrix} 3 & 2 & 3 \\ 2 & 2 & 1 \\ 2 & 1 & 1 \end{bmatrix}\begin{bmatrix} -\frac{1}{3} & -\frac{1}{3} & \frac{4}{3} \\ 0 & 1 & -1 \\ \frac{2}{3} & -\frac{1}{3} & -\frac{2}{3} \end{bmatrix} = \begin{bmatrix} 1 & 0 & 0 \\ 0 & 1 & 0 \\ 0 & 0 & 1 \end{bmatrix}$ and

$$\begin{bmatrix} -\frac{1}{3} & -\frac{1}{3} & \frac{4}{3} \\ 0 & 1 & -1 \\ \frac{2}{3} & -\frac{1}{3} & -\frac{2}{3} \end{bmatrix} \begin{bmatrix} 3 & 2 & 3 \\ 2 & 2 & 1 \\ 2 & 1 & 1 \end{bmatrix} = \begin{bmatrix} 1 & 0 & 0 \\ 0 & 1 & 0 \\ 0 & 0 & 1 \end{bmatrix}.$$

5. Using Formula (13), we find $A^{-1} = \dfrac{1}{(2)(3)-(1)(5)} \begin{bmatrix} 3 & -5 \\ -1 & 2 \end{bmatrix} = \begin{bmatrix} 3 & -5 \\ -1 & 2 \end{bmatrix}.$

7. Since $ad - bc = (3)(2) - (-2)(-3) = 6 - 6 = 0$, the inverse does not exist.

9.

$$\left[\begin{array}{ccc|ccc} 2 & -3 & -4 & 1 & 0 & 0 \\ 0 & 0 & -1 & 0 & 1 & 0 \\ 1 & -2 & 1 & 0 & 0 & 1 \end{array}\right] \xrightarrow{R_1 \leftrightarrow R_3} \left[\begin{array}{ccc|ccc} 1 & -2 & 1 & 0 & 0 & 1 \\ 0 & 0 & -1 & 0 & 1 & 0 \\ 2 & -3 & -4 & 1 & 0 & 0 \end{array}\right] \xrightarrow{R_3 - 2R_1}$$

$$\left[\begin{array}{ccc|ccc} 1 & -2 & 1 & 0 & 0 & 1 \\ 0 & 0 & -1 & 0 & 1 & 0 \\ 0 & 1 & -6 & 1 & 0 & -2 \end{array}\right] \xrightarrow{R_2 \leftrightarrow R_3} \left[\begin{array}{ccc|ccc} 1 & -2 & 1 & 0 & 0 & 1 \\ 0 & 1 & -6 & 1 & 0 & -2 \\ 0 & 0 & -1 & 0 & 1 & 0 \end{array}\right] \xrightarrow[\substack{-R_3}]{R_1 + 2R_2}$$

$$\left[\begin{array}{ccc|ccc} 1 & 0 & -11 & 2 & 0 & -3 \\ 0 & 1 & -6 & 1 & 0 & -2 \\ 0 & 0 & 1 & 0 & -1 & 0 \end{array}\right] \xrightarrow[\substack{R_2 + 6R_3}]{R_1 + 11R_3} \left[\begin{array}{ccc|ccc} 1 & 0 & 0 & 2 & -11 & -3 \\ 0 & 1 & 0 & 1 & -6 & -2 \\ 0 & 0 & 1 & 0 & -1 & 0 \end{array}\right].$$

Therefore, the required inverse is $\begin{bmatrix} 2 & -11 & -3 \\ 1 & -6 & -2 \\ 0 & -1 & 0 \end{bmatrix}.$

11.

$$\left[\begin{array}{ccc|ccc} 4 & 2 & 2 & 1 & 0 & 0 \\ -1 & -3 & 4 & 0 & 1 & 0 \\ 3 & -1 & 6 & 0 & 0 & 1 \end{array}\right] \xrightarrow{R_1 - R_3} \left[\begin{array}{ccc|ccc} 1 & 3 & -4 & 1 & 0 & -1 \\ -1 & -3 & 4 & 0 & 1 & 0 \\ 3 & -1 & 6 & 0 & 0 & 1 \end{array}\right]$$

$$\xrightarrow{R_2 + R_1} \left[\begin{array}{ccc|ccc} 1 & 3 & -4 & 1 & 0 & -1 \\ 0 & 0 & 0 & 1 & 1 & -1 \\ 3 & -1 & 6 & 0 & 0 & 1 \end{array}\right]$$

Because there is a row of zeros to the left of the vertical line, we see that the inverse does not exist.

13.
$$\begin{bmatrix} 1 & 4 & -1 & 1 & 0 & 0 \\ 2 & 3 & -2 & 0 & 1 & 0 \\ -1 & 2 & 3 & 0 & 0 & 1 \end{bmatrix} \xrightarrow{\substack{R_2 - 2R_1 \\ R_3 + R_1}} \begin{bmatrix} 1 & 4 & -1 & 1 & 0 & 0 \\ 0 & -5 & 0 & -2 & 1 & 0 \\ 0 & 6 & 2 & 1 & 0 & 1 \end{bmatrix} \xrightarrow{R_2 + R_3}$$

$$\begin{bmatrix} 1 & 4 & -1 & 1 & 0 & 0 \\ 0 & 1 & 2 & -1 & 1 & 1 \\ 0 & 6 & 2 & 1 & 0 & 1 \end{bmatrix} \xrightarrow{\substack{R_1 - 4R_2 \\ R_3 - 6R_2}} \begin{bmatrix} 1 & 0 & -9 & 5 & -4 & -4 \\ 0 & 1 & 2 & -1 & 1 & 1 \\ 0 & 0 & -10 & 7 & -6 & -5 \end{bmatrix} \xrightarrow{-\frac{1}{10}R_3}$$

$$\begin{bmatrix} 1 & 0 & -9 & 5 & -4 & -4 \\ 0 & 1 & 2 & -1 & 1 & 1 \\ 0 & 0 & 1 & -\frac{7}{10} & \frac{3}{5} & \frac{1}{2} \end{bmatrix} \xrightarrow{\substack{R_1 + 9R_3 \\ R_2 - 2R_3}} \begin{bmatrix} 1 & 0 & 0 & -\frac{13}{10} & \frac{7}{5} & \frac{1}{2} \\ 0 & 1 & 0 & \frac{2}{5} & -\frac{1}{5} & 0 \\ 0 & 0 & 1 & -\frac{7}{10} & \frac{3}{5} & \frac{1}{2} \end{bmatrix}$$

So $A^{-1} = \begin{bmatrix} -\frac{13}{10} & \frac{7}{5} & \frac{1}{2} \\ \frac{2}{5} & -\frac{1}{5} & 0 \\ -\frac{7}{10} & \frac{3}{5} & \frac{1}{2} \end{bmatrix}$.

15.
$$\begin{bmatrix} 1 & 1 & -1 & 1 & 1 & 0 & 0 & 0 \\ 2 & 1 & 1 & 0 & 0 & 1 & 0 & 0 \\ 2 & 1 & 0 & 1 & 0 & 0 & 1 & 0 \\ 2 & -1 & -1 & 3 & 0 & 0 & 0 & 1 \end{bmatrix} \xrightarrow{\substack{R_2 - 2R_1 \\ R_3 - 2R_1 \\ R_4 - 2R_1}} \begin{bmatrix} 1 & 1 & -1 & 1 & 1 & 0 & 0 & 0 \\ 0 & -1 & 3 & -2 & -2 & 1 & 0 & 0 \\ 0 & -1 & 2 & -1 & -2 & 0 & 1 & 0 \\ 0 & -3 & 1 & 1 & -2 & 0 & 0 & 1 \end{bmatrix} \xrightarrow{-R_2}$$

$$\begin{bmatrix} 1 & 1 & -1 & 1 & 1 & 0 & 0 & 0 \\ 0 & 1 & -3 & 2 & 2 & -1 & 0 & 0 \\ 0 & -1 & 2 & -1 & -2 & 0 & 1 & 0 \\ 0 & -3 & 1 & 1 & -2 & 0 & 0 & 1 \end{bmatrix} \xrightarrow{\substack{R_1 - R_2 \\ R_3 + R_2 \\ R_4 + 3R_2}} \begin{bmatrix} 1 & 0 & 2 & -1 & -1 & 1 & 0 & 0 \\ 0 & 1 & -3 & 2 & 2 & -1 & 0 & 0 \\ 0 & 0 & -1 & 1 & 0 & -1 & 1 & 0 \\ 0 & 0 & -8 & 7 & 4 & -3 & 0 & 1 \end{bmatrix} \xrightarrow{-R_3}$$

$$\begin{bmatrix} 1 & 0 & 2 & -1 & -1 & 1 & 0 & 0 \\ 0 & 1 & -3 & 2 & 2 & -1 & 0 & 0 \\ 0 & 0 & 1 & -1 & 0 & 1 & -1 & 0 \\ 0 & 0 & -8 & 7 & 4 & -3 & 0 & 1 \end{bmatrix} \xrightarrow{\substack{R_1 - 2R_3 \\ R_2 + 3R_3 \\ R_4 + 8R_3}} \begin{bmatrix} 1 & 0 & 0 & 1 & -1 & -1 & 2 & 0 \\ 0 & 1 & 0 & -1 & 2 & 2 & -3 & 0 \\ 0 & 0 & 1 & -1 & 0 & 1 & -1 & 0 \\ 0 & 0 & 0 & -1 & 4 & 5 & -8 & 1 \end{bmatrix}$$

$$\begin{array}{c} R_1+R_4 \\ R_2-R_4 \\ R_3-R_4 \\ \xrightarrow{\ -R_4\ } \end{array} \left[\begin{array}{cccc|cccc} 1 & 0 & 0 & 0 & 3 & 4 & -6 & 1 \\ 0 & 1 & 0 & 0 & -2 & -3 & 5 & -1 \\ 0 & 0 & 1 & 0 & -4 & -4 & 7 & -1 \\ 0 & 0 & 0 & 1 & -4 & -5 & 8 & -1 \end{array}\right].$$

So the required inverse is

$$A^{-1} = \begin{bmatrix} 3 & 4 & -6 & 1 \\ -2 & -3 & 5 & -1 \\ -4 & -4 & 7 & -1 \\ -4 & -5 & 8 & -1 \end{bmatrix}.$$

We can verify our result by showing that $A^{-1}A = A$. Thus,

$$\begin{bmatrix} 3 & 4 & -6 & 1 \\ -2 & -3 & 5 & -1 \\ -4 & -4 & 7 & -1 \\ -4 & -5 & 8 & -1 \end{bmatrix} \begin{bmatrix} 1 & 1 & -1 & 1 \\ 2 & 1 & 1 & 0 \\ 2 & 1 & 0 & 1 \\ 2 & -1 & -1 & 3 \end{bmatrix} = \begin{bmatrix} 1 & 0 & 0 & 0 \\ 0 & 1 & 0 & 0 \\ 0 & 0 & 1 & 0 \\ 0 & 0 & 0 & 1 \end{bmatrix}.$$

17.   a. $A = \begin{bmatrix} 2 & 5 \\ 1 & 3 \end{bmatrix}$, $X = \begin{bmatrix} x \\ y \end{bmatrix}$, $B = \begin{bmatrix} 3 \\ 2 \end{bmatrix}$;

b. $X = A^{-1}B = \begin{bmatrix} 3 & -5 \\ -1 & 2 \end{bmatrix} \begin{bmatrix} 3 \\ 2 \end{bmatrix} = \begin{bmatrix} -1 \\ 1 \end{bmatrix}$;

19.   a. $A = \begin{bmatrix} 2 & -3 & -4 \\ 0 & 0 & -1 \\ 1 & -2 & 1 \end{bmatrix}$, $X = \begin{bmatrix} x \\ y \\ z \end{bmatrix}$, $B = \begin{bmatrix} 4 \\ 3 \\ -8 \end{bmatrix}$

$$X = A^{-1}B = \begin{bmatrix} 2 & -11 & -3 \\ 1 & -6 & -2 \\ 0 & -1 & 0 \end{bmatrix} \begin{bmatrix} 4 \\ 3 \\ -8 \end{bmatrix} = \begin{bmatrix} -1 \\ 2 \\ -3 \end{bmatrix}$$

21. a. $A = \begin{bmatrix} 1 & 4 & -1 \\ 2 & 3 & -2 \\ -1 & 2 & 3 \end{bmatrix}$, $X = \begin{bmatrix} x \\ y \\ z \end{bmatrix}$, $B = \begin{bmatrix} 3 \\ 1 \\ 7 \end{bmatrix}$; b. $X = A^{-1}B = \begin{bmatrix} -\frac{13}{10} & \frac{7}{5} & \frac{1}{2} \\ \frac{2}{5} & -\frac{1}{5} & 0 \\ -\frac{7}{10} & \frac{3}{5} & \frac{1}{2} \end{bmatrix} \begin{bmatrix} 3 \\ 1 \\ 7 \end{bmatrix} = \begin{bmatrix} 1 \\ 1 \\ 2 \end{bmatrix}$.

23. a. $A = \begin{bmatrix} 1 & 1 & -1 & 1 \\ 2 & 1 & 1 & 0 \\ 2 & 1 & 0 & 1 \\ 2 & -1 & -1 & 3 \end{bmatrix}$, $X = \begin{bmatrix} x_1 \\ x_2 \\ x_3 \\ x_4 \end{bmatrix}$, $B = \begin{bmatrix} 6 \\ 4 \\ 7 \\ 9 \end{bmatrix}$.

b. $X = A^{-1}B = \begin{bmatrix} 3 & 4 & -6 & 1 \\ -2 & -3 & 5 & -1 \\ -4 & -4 & 7 & -1 \\ -4 & -5 & 8 & -1 \end{bmatrix} \begin{bmatrix} 6 \\ 4 \\ 7 \\ 9 \end{bmatrix} = \begin{bmatrix} 1 \\ 2 \\ 0 \\ 3 \end{bmatrix}$.

25. a. $A = \begin{bmatrix} 1 & 2 \\ 2 & -1 \end{bmatrix}$, $X = \begin{bmatrix} x \\ y \end{bmatrix}$, $B = \begin{bmatrix} b_1 \\ b_2 \end{bmatrix}$;

b (i). $X = A^{-1}B = \begin{bmatrix} 0.2 & 0.4 \\ 0.4 & -0.2 \end{bmatrix} \begin{bmatrix} 14 \\ 5 \end{bmatrix} = \begin{bmatrix} 4.8 \\ 4.6 \end{bmatrix}$ and we conclude that $x = 4.8$ and $y = 4.6$.

(ii). $X = A^{-1}B = \begin{bmatrix} 0.2 & 0.4 \\ 0.4 & -0.2 \end{bmatrix} \begin{bmatrix} 4 \\ -1 \end{bmatrix} = \begin{bmatrix} 0.4 \\ 1.8 \end{bmatrix}$ and we conclude that $x = 0.4$ and $y = 1.8$.

27. a. First we find A^{-1}.

$$\begin{bmatrix} 1 & 2 & 1 & | & 1 & 0 & 0 \\ 1 & 1 & 1 & | & 0 & 1 & 0 \\ 3 & 1 & 1 & | & 0 & 0 & 1 \end{bmatrix} \xrightarrow[R_3 - 3R_1]{R_2 - R_1} \begin{bmatrix} 1 & 2 & 1 & | & 1 & 0 & 0 \\ 0 & -1 & 0 & | & -1 & 1 & 0 \\ 0 & -5 & -2 & | & -3 & 0 & 1 \end{bmatrix} \xrightarrow{-R_2}$$

$$\begin{bmatrix} 1 & 2 & 1 & | & 1 & 0 & 0 \\ 0 & 1 & 0 & | & 1 & -1 & 0 \\ 0 & -5 & -2 & | & -3 & 0 & 1 \end{bmatrix} \xrightarrow[R_3 + 5R_2]{R_1 - 2R_2} \begin{bmatrix} 1 & 0 & 1 & | & -1 & 2 & 0 \\ 0 & 1 & 0 & | & 1 & -1 & 0 \\ 0 & 0 & -2 & | & 2 & -5 & 1 \end{bmatrix} \xrightarrow{-\frac{1}{2}R_3}$$

$$\begin{bmatrix} 1 & 0 & 1 & -1 & 2 & 0 \\ 0 & 1 & 0 & 1 & -1 & 0 \\ 0 & 0 & 1 & -1 & \frac{5}{2} & -\frac{1}{2} \end{bmatrix} \xrightarrow{R_1 - R_3} \begin{bmatrix} 1 & 0 & 0 & 0 & -\frac{1}{2} & \frac{1}{2} \\ 0 & 1 & 0 & 1 & -1 & 0 \\ 0 & 0 & 1 & -1 & \frac{5}{2} & -\frac{1}{2} \end{bmatrix}$$

$$\begin{bmatrix} 1 & 2 & 1 \\ 1 & 1 & 1 \\ 3 & 1 & 1 \end{bmatrix} \begin{bmatrix} x \\ y \\ z \end{bmatrix} = \begin{bmatrix} b_1 \\ b_2 \\ b_3 \end{bmatrix}$$

b. (i).
$$\begin{bmatrix} x \\ y \\ z \end{bmatrix} = \begin{bmatrix} 0 & -\frac{1}{2} & \frac{1}{2} \\ 1 & -1 & 0 \\ -1 & \frac{5}{2} & -\frac{1}{2} \end{bmatrix} \begin{bmatrix} 7 \\ 4 \\ 2 \end{bmatrix} = \begin{bmatrix} -1 \\ 3 \\ 2 \end{bmatrix}$$
and we conclude that $x = -1$, $y = 3$, and $z = 2$

(ii).
$$\begin{bmatrix} x \\ y \\ z \end{bmatrix} = \begin{bmatrix} 0 & -\frac{1}{2} & \frac{1}{2} \\ 1 & -1 & 0 \\ -1 & \frac{5}{2} & -\frac{1}{2} \end{bmatrix} \begin{bmatrix} 5 \\ -3 \\ -1 \end{bmatrix} = \begin{bmatrix} 1 \\ 8 \\ -12 \end{bmatrix}$$
and we conclude that $x = 1$, $y = 8$, and $z = -12$.

29. a.
$$\begin{bmatrix} 3 & 2 & -1 & 1 & 0 & 0 \\ 2 & -3 & 1 & 0 & 1 & 0 \\ 1 & -1 & -1 & 0 & 0 & 1 \end{bmatrix} \xrightarrow{R_1 \leftrightarrow R_3} \begin{bmatrix} 1 & -1 & -1 & 0 & 0 & 1 \\ 2 & -3 & 1 & 0 & 1 & 0 \\ 3 & 2 & -1 & 1 & 0 & 0 \end{bmatrix} \xrightarrow[R_3 - 3R_1]{R_2 - 2R_1}$$

$$\begin{bmatrix} 1 & -1 & -1 & 0 & 0 & 1 \\ 0 & -1 & 3 & 0 & 1 & -2 \\ 0 & 5 & 2 & 1 & 0 & -3 \end{bmatrix} \xrightarrow{-R_2} \begin{bmatrix} 1 & -1 & -1 & 0 & 0 & 1 \\ 0 & 1 & -3 & 0 & -1 & 2 \\ 0 & 5 & 2 & 1 & 0 & -3 \end{bmatrix} \xrightarrow[R_3 - 5R_2]{R_1 + R_2}$$

$$\begin{bmatrix} 1 & 0 & -4 & 0 & -1 & 3 \\ 0 & 1 & -3 & 0 & -1 & 2 \\ 0 & 0 & 17 & 1 & 5 & -13 \end{bmatrix} \xrightarrow{\frac{1}{17}R_3} \begin{bmatrix} 1 & 0 & -4 & 0 & -1 & 3 \\ 0 & 1 & -3 & 0 & -1 & 2 \\ 0 & 0 & 1 & \frac{1}{17} & \frac{5}{17} & -\frac{13}{17} \end{bmatrix}$$

$$\xrightarrow[R_2+3R_3]{R_1+4R_3} \begin{bmatrix} 1 & 0 & 0 \\ 0 & 1 & 0 \\ 0 & 0 & 1 \end{bmatrix} \begin{array}{|ccc} \frac{4}{17} & \frac{3}{17} & -\frac{1}{17} \\ \frac{3}{17} & -\frac{2}{17} & -\frac{5}{17} \\ \frac{1}{17} & \frac{5}{17} & -\frac{13}{17} \end{array} . \quad \text{Therefore } A^{-1} = \begin{bmatrix} \frac{4}{17} & \frac{3}{17} & -\frac{1}{17} \\ \frac{3}{17} & -\frac{2}{17} & -\frac{5}{17} \\ \frac{1}{17} & \frac{5}{17} & -\frac{13}{17} \end{bmatrix} .$$

Next, $\begin{bmatrix} 3 & 2 & -1 \\ 2 & -3 & 1 \\ 1 & -1 & -1 \end{bmatrix} \begin{bmatrix} x \\ y \\ z \end{bmatrix} = \begin{bmatrix} b_1 \\ b_2 \\ b_3 \end{bmatrix}$

b. (i) $\begin{bmatrix} x \\ y \\ z \end{bmatrix} = \begin{bmatrix} \frac{4}{17} & \frac{3}{17} & -\frac{1}{17} \\ \frac{3}{17} & -\frac{2}{17} & -\frac{5}{17} \\ \frac{1}{17} & \frac{5}{17} & -\frac{13}{17} \end{bmatrix} \begin{bmatrix} 2 \\ -2 \\ 4 \end{bmatrix} = \begin{bmatrix} -\frac{2}{17} \\ -\frac{10}{17} \\ -\frac{60}{17} \end{bmatrix}$

We conclude that $x = -2/17$, $y = -10/17$, and $z = -60/17$.

(ii) $\begin{bmatrix} x \\ y \\ z \end{bmatrix} = \begin{bmatrix} \frac{4}{17} & \frac{3}{17} & -\frac{1}{17} \\ \frac{3}{17} & -\frac{2}{17} & -\frac{5}{17} \\ \frac{1}{17} & \frac{5}{17} & -\frac{13}{17} \end{bmatrix} \begin{bmatrix} 8 \\ -3 \\ 6 \end{bmatrix} = \begin{bmatrix} 1 \\ 0 \\ -5 \end{bmatrix}$. We conclude that $x = 1$, $y = 0$, and $z = -5$.

31. a. $AX = B_1$ and $AX = B_2$, where

$$A = \begin{bmatrix} 1 & 1 & 1 & 1 \\ 1 & -1 & -1 & 1 \\ 0 & 1 & 2 & 2 \\ 1 & 2 & 1 & -2 \end{bmatrix}, \ X = \begin{bmatrix} x_1 \\ x_2 \\ x_3 \\ x_4 \end{bmatrix}, \ B_1 = \begin{bmatrix} 1 \\ -1 \\ 4 \\ 0 \end{bmatrix} \text{ and } B_2 = \begin{bmatrix} 2 \\ 8 \\ 4 \\ -1 \end{bmatrix}.$$

We first find A^{-1}.

$$\begin{bmatrix} 1 & 1 & 1 & 1 \\ 1 & -1 & -1 & 1 \\ 0 & 1 & 2 & 2 \\ 1 & 2 & 1 & -2 \end{bmatrix} \begin{array}{|cccc} 1 & 0 & 0 & 0 \\ 0 & 1 & 0 & 0 \\ 0 & 0 & 1 & 0 \\ 0 & 0 & 0 & 1 \end{array} \xrightarrow[R_4-R_1]{R_2-R_1} \begin{bmatrix} 1 & 1 & 1 & 1 \\ 0 & -2 & -2 & 0 \\ 0 & 1 & 2 & 2 \\ 0 & 1 & 0 & -3 \end{bmatrix} \begin{array}{|cccc} 1 & 0 & 0 & 0 \\ -1 & 1 & 0 & 0 \\ 0 & 0 & 1 & 0 \\ -1 & 0 & 0 & 1 \end{array}$$

$$\xrightarrow{R_2 \leftrightarrow R_3} \begin{bmatrix} 1 & 1 & 1 & 1 & | & 1 & 0 & 0 & 0 \\ 0 & 1 & 2 & 2 & | & 0 & 0 & 1 & 0 \\ 0 & -2 & -2 & 0 & | & -1 & 1 & 0 & 0 \\ 0 & 1 & 0 & -3 & | & -1 & 0 & 0 & 1 \end{bmatrix} \begin{matrix} R_1 - R_2 \\ R_3 + 2R_2 \\ R_4 - R_2 \end{matrix} \longrightarrow$$

$$\begin{bmatrix} 1 & 0 & -1 & -1 & | & 1 & 0 & -1 & 0 \\ 0 & 1 & 2 & 2 & | & 0 & 0 & 1 & 0 \\ 0 & 0 & 2 & 4 & | & -1 & 1 & 2 & 0 \\ 0 & 0 & -2 & -5 & | & -1 & 0 & -1 & 1 \end{bmatrix} \xrightarrow{\frac{1}{2}R_3} \begin{bmatrix} 1 & 0 & -1 & -1 & | & 1 & 0 & -1 & 0 \\ 0 & 1 & 2 & 2 & | & 0 & 0 & 1 & 0 \\ 0 & 0 & 1 & 2 & | & -\frac{1}{2} & \frac{1}{2} & 1 & 0 \\ 0 & 0 & -2 & -5 & | & -1 & 0 & -1 & 1 \end{bmatrix}$$

$$\begin{matrix} R_1 + R_3 \\ R_2 - 2R_3 \\ R_4 + 2R_3 \end{matrix} \longrightarrow \begin{bmatrix} 1 & 0 & 0 & 1 & | & \frac{1}{2} & \frac{1}{2} & 0 & 0 \\ 0 & 1 & 0 & -2 & | & 1 & -1 & -1 & 0 \\ 0 & 0 & 1 & 2 & | & -\frac{1}{2} & \frac{1}{2} & 1 & 0 \\ 0 & 0 & 0 & -1 & | & -2 & 1 & 1 & 1 \end{bmatrix} \begin{matrix} R_1 + R_4 \\ R_2 - 2R_4 \\ R_3 + 2R_4 \\ -R_4 \end{matrix} \longrightarrow$$

$$\begin{bmatrix} 1 & 0 & 0 & 0 & | & -\frac{3}{2} & \frac{3}{2} & 1 & 1 \\ 0 & 1 & 0 & 0 & | & 5 & -3 & -3 & -2 \\ 0 & 0 & 1 & 0 & | & -\frac{9}{2} & \frac{5}{2} & 3 & 2 \\ 0 & 0 & 0 & 1 & | & 2 & -1 & -1 & -1 \end{bmatrix}. \quad \text{So} \quad A^{-1} = \begin{bmatrix} -\frac{3}{2} & \frac{3}{2} & 1 & 1 \\ 5 & -3 & -3 & -2 \\ -\frac{9}{2} & \frac{5}{2} & 3 & 2 \\ 2 & -1 & -1 & -1 \end{bmatrix}.$$

b. (i).
$$\begin{bmatrix} x_1 \\ x_2 \\ x_3 \\ x_4 \end{bmatrix} = \begin{bmatrix} -\frac{3}{2} & \frac{3}{2} & 1 & 1 \\ 5 & -3 & -3 & -2 \\ -\frac{9}{2} & \frac{5}{2} & 3 & 2 \\ 2 & -1 & -1 & -1 \end{bmatrix} \begin{bmatrix} 1 \\ -1 \\ 4 \\ 0 \end{bmatrix} = \begin{bmatrix} 1 \\ -4 \\ 5 \\ -1 \end{bmatrix}$$

and we conclude that $x_1 = 1$, $x_2 = -4$, $x_3 = 5$, and $x_4 = -1$.

(ii).
$$\begin{bmatrix} x_1 \\ x_2 \\ x_3 \\ x_4 \end{bmatrix} = \begin{bmatrix} -\frac{3}{2} & \frac{3}{2} & 1 & 1 \\ 5 & -3 & -3 & -2 \\ -\frac{9}{2} & \frac{5}{2} & 3 & 2 \\ 2 & -1 & -1 & -1 \end{bmatrix} \begin{bmatrix} 2 \\ 8 \\ 4 \\ -1 \end{bmatrix} = \begin{bmatrix} 12 \\ -24 \\ 21 \\ -7 \end{bmatrix}$$

and we conclude that $x_1 = 12$, $x_2 = -24$, $x_3 = 21$, and $x_4 = -7$.

33. **a.** Using Formula (13), we find $A^{-1} = \dfrac{1}{(2)(-5)-(-4)(3)}\begin{bmatrix} -5 & -3 \\ 4 & 2 \end{bmatrix} = \begin{bmatrix} -\frac{5}{2} & -\frac{3}{2} \\ 2 & 1 \end{bmatrix}.$

b. Using Formula (13) once again, we find

$$\left(A^{-1}\right)^{-1} = \frac{1}{(-\frac{5}{2})(1) - 2(-\frac{3}{2})}\begin{bmatrix} 1 & \frac{3}{2} \\ -2 & -\frac{5}{2} \end{bmatrix} = \begin{bmatrix} 2 & 3 \\ -4 & -5 \end{bmatrix} = A.$$

35. **a.** $ABC = \begin{bmatrix} 2 & -5 \\ 1 & -3 \end{bmatrix}\begin{bmatrix} 4 & 3 \\ 1 & 1 \end{bmatrix}\begin{bmatrix} 2 & 3 \\ -2 & 1 \end{bmatrix} = \begin{bmatrix} 2 & -5 \\ 1 & -3 \end{bmatrix}\begin{bmatrix} 2 & 15 \\ 0 & 4 \end{bmatrix} = \begin{bmatrix} 4 & 10 \\ 2 & 3 \end{bmatrix}.$

Using the formula for finding the inverse of a 2×2 matrix, we find

$$A^{-1} = \begin{bmatrix} 3 & -5 \\ 1 & -2 \end{bmatrix}, \quad B^{-1} = \begin{bmatrix} 1 & -3 \\ -1 & 4 \end{bmatrix}, \quad C^{-1} = \begin{bmatrix} \frac{1}{8} & -\frac{3}{8} \\ \frac{1}{4} & \frac{1}{4} \end{bmatrix}.$$

b. Using the formula for finding the inverse of a 2×2 matrix, we find

$$(ABC)^{-1} = \begin{bmatrix} -\frac{3}{8} & \frac{5}{4} \\ \frac{1}{4} & -\frac{1}{2} \end{bmatrix}$$

$$C^{-1}B^{-1}A^{-1} = \begin{bmatrix} \frac{1}{8} & -\frac{3}{8} \\ \frac{1}{4} & \frac{1}{4} \end{bmatrix}\begin{bmatrix} 1 & -3 \\ -1 & 4 \end{bmatrix}\begin{bmatrix} 3 & -5 \\ 1 & -2 \end{bmatrix}$$

$$= \begin{bmatrix} \frac{1}{8} & -\frac{3}{8} \\ \frac{1}{4} & \frac{1}{4} \end{bmatrix}\begin{bmatrix} 0 & 1 \\ 1 & -3 \end{bmatrix} = \begin{bmatrix} -\frac{3}{8} & \frac{5}{4} \\ \frac{1}{4} & -\frac{1}{2} \end{bmatrix}.$$

Therefore, $(ABC)^{-1} = C^{-1}B^{-1}A^{-1}.$

37. Let x denote the number of copies of the deluxe edition and y the number of copies of the standard edition demanded per month when the unit prices are p and q dollars, respectively. Then the three systems of linear equations

$$5x + y = 20000 \qquad 5x + y = 25000 \qquad 5x + y = 25000$$
$$x + 3y = 15000 \qquad x + 3y = 15000 \qquad x + 3y = 20000$$

give the quantity demanded of each edition at the stated price. These systems may be written in the form $AX = B_1$, $AX = B_2$, and $AX = B_3$, where

$$A = \begin{bmatrix} 5 & 1 \\ 1 & 3 \end{bmatrix}, \quad B_1 = \begin{bmatrix} 20000 \\ 15000 \end{bmatrix}, \quad B_2 = \begin{bmatrix} 25000 \\ 15000 \end{bmatrix}, \quad \text{and} \quad B_3 = \begin{bmatrix} 25000 \\ 20000 \end{bmatrix}$$

Using the formula for finding the inverse of a 2×2 matrix, with $a = 5, b = 1, c = 1$, $d = 3$, and $D = ad - bc = (5)(3) - (1)(1) = 14$, we find that $A^{-1} = \begin{bmatrix} \frac{3}{14} & -\frac{1}{14} \\ -\frac{1}{14} & \frac{5}{14} \end{bmatrix}$.

a. $\begin{bmatrix} x \\ y \end{bmatrix} = \begin{bmatrix} \frac{3}{14} & -\frac{1}{14} \\ -\frac{1}{14} & \frac{5}{14} \end{bmatrix} \begin{bmatrix} 20,000 \\ 15,000 \end{bmatrix} = \begin{bmatrix} 3,214 \\ 3,929 \end{bmatrix}$
b. $\begin{bmatrix} x \\ y \end{bmatrix} = \begin{bmatrix} \frac{3}{14} & -\frac{1}{14} \\ -\frac{1}{14} & \frac{5}{14} \end{bmatrix} \begin{bmatrix} 25,000 \\ 15,000 \end{bmatrix} = \begin{bmatrix} 4,286 \\ 3,571 \end{bmatrix}$

c. $\begin{bmatrix} x \\ y \end{bmatrix} = \begin{bmatrix} \frac{3}{14} & -\frac{1}{14} \\ -\frac{1}{14} & \frac{5}{14} \end{bmatrix} \begin{bmatrix} 25,000 \\ 20,000 \end{bmatrix} = \begin{bmatrix} 3,929 \\ 5,357 \end{bmatrix}$.

39. Let x, y, and z denote the number of acres of soybeans, corn, and wheat to be cultivated, respectively. Furthermore, let a, b, and c denote the amount of land available; the amount of labor available, and the amount of money available for seeds, respectively. Then we have the system

$$x + y + z = a \qquad \text{(land)}$$
$$2x + 6y + 6z = b \qquad \text{(labor)}$$
$$12x + 20y + 8z = c \qquad \text{(seeds)}$$

The system can be written in the form $AX = B$, where

$$A = \begin{bmatrix} 1 & 1 & 1 \\ 2 & 6 & 6 \\ 12 & 20 & 8 \end{bmatrix}, \quad X = \begin{bmatrix} x \\ y \\ z \end{bmatrix}, \quad \text{and} \quad B = \begin{bmatrix} a \\ b \\ c \end{bmatrix}$$

Using the technique for finding A^{-1} developed in this section, we find

$$A^{-1} = \begin{bmatrix} \frac{3}{2} & -\frac{1}{4} & 0 \\ -\frac{7}{6} & \frac{1}{12} & \frac{1}{12} \\ \frac{2}{3} & \frac{1}{6} & -\frac{1}{12} \end{bmatrix}.$$

a. Here $a = 1000$, $b = 4400$, and $c = 13,200$. Therefore

$$X = A^{-1}B = \begin{bmatrix} \frac{3}{2} & -\frac{1}{4} & 0 \\ -\frac{7}{6} & \frac{1}{12} & \frac{1}{12} \\ \frac{2}{3} & \frac{1}{6} & -\frac{1}{12} \end{bmatrix} \begin{bmatrix} 1,000 \\ 4,400 \\ 13,200 \end{bmatrix} = \begin{bmatrix} 400 \\ 300 \\ 300 \end{bmatrix}$$

So, Jackson Farms should cultivate 400, 300, and 300 acres of soybeans, corn, and wheat, respectively.

b. Here $a = 1200$, $b = 5200$, and $c = 16,400$. Therefore,

$$X = A^{-1}B = \begin{bmatrix} \frac{3}{2} & -\frac{1}{4} & 0 \\ -\frac{7}{6} & \frac{1}{12} & \frac{1}{12} \\ \frac{2}{3} & \frac{1}{6} & -\frac{1}{12} \end{bmatrix} \begin{bmatrix} 1,200 \\ 5,200 \\ 16,400 \end{bmatrix} = \begin{bmatrix} 500 \\ 400 \\ 300 \end{bmatrix}$$

So, Jackson Farms should cultivate 500, 400, and 300 acres of soybeans, corn, and wheat, respectively.

41. Let x, y, and z denote the amount to be invested in high-risk, medium-risk, and low-risk stocks, respectively. Next, let a denote the amount to be invested and let c denote the return on the investments. Then, we have the system

$$\begin{aligned} x + \quad y + \quad z &= a \\ x + \quad y - \quad z &= 0 \qquad \text{(since } z = x + y) \\ 0.15x + 0.1y + 0.06z &= c \end{aligned}$$

The system is equivalent to the matrix equation $AX = B$, where

$$A = \begin{bmatrix} 1 & 1 & 1 \\ 1 & 1 & -1 \\ .15 & .10 & .06 \end{bmatrix}, \quad X = \begin{bmatrix} x \\ y \\ z \end{bmatrix}, \quad \text{and} \quad B = \begin{bmatrix} a \\ 0 \\ c \end{bmatrix}.$$

We find $A^{-1} = \begin{bmatrix} -1.6 & -0.4 & 20 \\ 2.1 & 0.9 & -20 \\ 0.5 & -0.5 & 0 \end{bmatrix}$.

a. Here $a = 200,000$ and $c = 20,000$. Therefore,

$$X = A^{-1}B = \begin{bmatrix} -1.6 & -0.4 & 20 \\ 2.1 & 0.9 & -20 \\ 0.5 & -0.5 & 0 \end{bmatrix} \begin{bmatrix} 200,000 \\ 0 \\ 20,000 \end{bmatrix} = \begin{bmatrix} 80,000 \\ 20,000 \\ 100,000 \end{bmatrix}.$$

So, the club should invest $80,000 in high-risk, $20,000 in medium risk, and $100,000 in low risk stocks.

b. Here $a = 220,000$ and $c = 22,000$. The solution is $x = 88,000$, $y = 22,000$, and $z = 110,000$; that is, the club should invest $88,000 in high-risk, $22,000 in medium-risk, and $110,000 in low-risk stocks.

c. Here $a = 240,000$ and $c = 22,000$. The result is $56,000 in high-risk stocks, $64,000 in medium-risk stocks, and $120,000 in low-risk stocks.

43. True. Multiplying both sides of the equation by cA yields

$$I = (cA)(cA)^{-1} = (cA)\left[\frac{1}{c}(A^{-1})\right] = c\left(\frac{1}{c}\right)AA^{-1} = I .$$

45. True. $AX = B$ can have a unique solution only if A^{-1} exists, in which case the solution is found as follows:

$$A^{-1}(AX) = A^{-1}B$$
$$(A^{-1}A)X = A^{-1}B$$
$$IX = A^{-1}B$$
$$X = A^{-1}B$$

USING TECHNOLOGY EXERCISES 2.6, page 154

1. $\begin{bmatrix} 0.36 & 0.04 & -0.36 \\ 0.06 & 0.05 & 0.20 \\ -0.19 & 0.10 & 0.09 \end{bmatrix}$

3. $\begin{bmatrix} 0.01 & -0.09 & 0.31 & -0.11 \\ -0.25 & 0.58 & -0.15 & -0.02 \\ 0.86 & -0.42 & 0.07 & -0.37 \\ -0.27 & 0.01 & -0.05 & 0.31 \end{bmatrix}$

5. $\begin{bmatrix} 0.30 & 0.85 & -0.10 & -0.77 & -0.11 \\ -0.21 & 0.10 & 0.01 & -0.26 & 0.21 \\ 0.03 & -0.16 & 0.12 & -0.01 & 0.03 \\ -0.14 & -0.46 & 0.13 & 0.71 & -0.05 \\ 0.10 & -0.05 & -0.10 & -0.03 & 0.11 \end{bmatrix}$

7. $x = 1.2$, $y = 3.6$, and $z = 2.7$.   9. $x_1 = 2.50$, $x_2 = -0.88$, $x_3 = 0.70$, and $x_4 = 0.51$.

EXERCISES 2.7, page 161

1.   a. The amount of agricultural products consumed in the production of $100 million

worth of manufactured goods is given by $(100)(0.10)$, or $10 million.

b. The amount of manufactured goods required to produce $200 million of all goods in the economy is given by $200(0.1 + 0.4 + 0.3) = 160$, or $160 million.

c. From the input-output matrix, we see that the agricultural sector consumes the greatest amount of agricultural products, namely, 0.4 units, in the production of each unit of goods in that sector. The manufacturing and transportation sectors consume the least, 0.1 units each.

3. Multiplying both sides of the given equation on the left by $(I - A)^{-1}$, we see that
$X = (I - A)^{-1}D.$

Now, $(I - A) = \begin{bmatrix} 1 & 0 \\ 0 & 1 \end{bmatrix} - \begin{bmatrix} 0.4 & 0.2 \\ 0.3 & 0.1 \end{bmatrix} = \begin{bmatrix} 0.6 & -0.2 \\ -0.3 & 0.9 \end{bmatrix}.$

Using the formula for finding the inverse of a 2×2 matrix, we find

$(I - A)^{-1} = \begin{bmatrix} 1.875 & 0.417 \\ 0.625 & 1.25 \end{bmatrix}.$

Then, $(I - A)^{-1}X = \begin{bmatrix} 1.875 & 0.417 \\ 0.625 & 1.25 \end{bmatrix}\begin{bmatrix} 10 \\ 12 \end{bmatrix} = \begin{bmatrix} 23.754 \\ 21.25 \end{bmatrix}.$

5. We first compute $(I - A) = \begin{bmatrix} 1 & 0 \\ 0 & 1 \end{bmatrix} - \begin{bmatrix} 0.5 & 0.2 \\ 0.2 & 0.5 \end{bmatrix} = \begin{bmatrix} 0.5 & -0.2 \\ -0.2 & 0.5 \end{bmatrix}$

Using the formula for finding the inverse of a 2×2 matrix, we find

$(I - A)^{-1} = \begin{bmatrix} 2.381 & 0.952 \\ 0.952 & 2.381 \end{bmatrix}.$

Then $\begin{bmatrix} x \\ y \end{bmatrix} = \begin{bmatrix} 2.381 & 0.952 \\ 0.952 & 2.381 \end{bmatrix}\begin{bmatrix} 10 \\ 20 \end{bmatrix} = \begin{bmatrix} 42.85 \\ 57.14 \end{bmatrix}$

7. We verify
$$(I - A)(I - A)^{-1} = \begin{bmatrix} 0.92 & -0.60 & -0.30 \\ -0.04 & 0.98 & -0.01 \\ -0.02 & 0 & 0.94 \end{bmatrix}\begin{bmatrix} 1.13 & 0.69 & 0.37 \\ 0.05 & 1.05 & 0.03 \\ 0.02 & 0.02 & 1.07 \end{bmatrix} = \begin{bmatrix} 1 & 0 & 0 \\ 0 & 1 & 0 \\ 0 & 0 & 1 \end{bmatrix}$$

9. a. $A = \begin{bmatrix} 0.2 & 0.4 \\ 0.3 & 0.3 \end{bmatrix}$ and

$$(I - A) = \begin{bmatrix} 1 & 0 \\ 0 & 1 \end{bmatrix} - \begin{bmatrix} 0.2 & 0.4 \\ 0.3 & 0.3 \end{bmatrix} = \begin{bmatrix} 0.8 & -0.4 \\ -0.3 & 0.7 \end{bmatrix}.$$

Using the formula for finding the inverse of a 2×2 matrix, we find

$$(I - A)^{-1} = \begin{bmatrix} 1.591 & 0.909 \\ 0.682 & 1.818 \end{bmatrix}.$$

Then $\begin{bmatrix} x \\ y \end{bmatrix} = \begin{bmatrix} 1.591 & 0.909 \\ 0.682 & 1.818 \end{bmatrix} \begin{bmatrix} 120 \\ 140 \end{bmatrix} = \begin{bmatrix} 318.18 \\ 336.36 \end{bmatrix}$

To fullfill consumer demand, \$318.2 million worth of agricultural goods and \$336.4 million worth of manufactured goods should be produced.

b. The net value of goods consumed in the internal process of production is

$$AX = X - D = \begin{bmatrix} 318.18 \\ 336.36 \end{bmatrix} - \begin{bmatrix} 120 \\ 140 \end{bmatrix} = \begin{bmatrix} 198.18 \\ 196.36 \end{bmatrix}.$$

or \$198.2 million of agricultural goods and \$196.4 million worth of manufactured goods.

11. a.

$$(I - A) = \begin{bmatrix} 1 & 0 & 0 \\ 0 & 1 & 0 \\ 0 & 0 & 1 \end{bmatrix} - \begin{bmatrix} 0.4 & 0.1 & 0.1 \\ 0.1 & 0.4 & 0.3 \\ 0.2 & 0.2 & 0.2 \end{bmatrix} = \begin{bmatrix} 0.6 & -0.1 & -0.1 \\ -0.1 & 0.6 & -0.3 \\ -0.2 & -0.2 & 0.8 \end{bmatrix}$$

Using the methods of Section 2.6 we next compute the inverse of $(1 - A)^{-1}$ and use this value to find

$$X = (1 - A)^{-1} D = \begin{bmatrix} 1.875 & 0.446 & 0.402 \\ 0.625 & 2.054 & 0.848 \\ 0.625 & 0.625 & 1.563 \end{bmatrix} \begin{bmatrix} 200 \\ 100 \\ 60 \end{bmatrix} = \begin{bmatrix} 443.7 \\ 381.3 \\ 281.3 \end{bmatrix}.$$

Therefore, to fulfull demand, \$443.7 million worth of agricultural products, \$381.3 million worth of manufactured products, and \$281.3 million worth of transportation services should be produced.

b. To meet the gross output, the value of goods and transportation consumed in the internal process of production is

$$AX = X - D = \begin{bmatrix} 443.7 \\ 381.3 \\ 281.3 \end{bmatrix} - \begin{bmatrix} 200 \\ 100 \\ 60 \end{bmatrix} = \begin{bmatrix} 243.7 \\ 281.3 \\ 221.3 \end{bmatrix},$$

or \$243.7 million worth of agricultural products, \$281.3 million worth of manufactured services, and $221.3 million worth of transportation services.

13. We want to solve the equation $(I - A)X = D$ for X, the total output matrix. First, we compute

$$(I - A) = \begin{bmatrix} 1 & 0 \\ 0 & 1 \end{bmatrix} - \begin{bmatrix} 0.4 & 0.2 \\ 0.3 & 0.5 \end{bmatrix} = \begin{bmatrix} 0.6 & -0.2 \\ -0.3 & 0.5 \end{bmatrix}.$$

Using the formula for finding the inverse of a 2×2 matrix, we find

$$(I - A)^{-1} = \begin{bmatrix} 2.08 & 0.833 \\ 1.25 & 2.5 \end{bmatrix}$$

Therefore,

$$X = (I - A)^{-1} D = \begin{bmatrix} 2.08 & 0.833 \\ 1.25 & 2.5 \end{bmatrix} \begin{bmatrix} 12 \\ 24 \end{bmatrix} = \begin{bmatrix} 45 \\ 75 \end{bmatrix}.$$

We conclude that $45 million worth of goods of one industry and $75 million worth of goods of the other industry must be produced.

15. First, we compute

$$I - A = \begin{bmatrix} 1 & 0 & 0 \\ 0 & 1 & 0 \\ 0 & 0 & 1 \end{bmatrix} - \begin{bmatrix} 0.2 & 0.4 & 0.2 \\ 0.5 & 0 & 0.5 \\ 0 & 0.2 & 0 \end{bmatrix} = \begin{bmatrix} 0.8 & -0.4 & -0.2 \\ -0.5 & 1 & -0.5 \\ 0 & -0.2 & 1 \end{bmatrix}.$$

Next, using the Gauss-Jordan method, we find

$$(I - A)^{-1} = \begin{bmatrix} 1.8 & 0.88 & 0.80 \\ 1 & 1.6 & 1 \\ 0.2 & 0.32 & 1.20 \end{bmatrix}$$

Then $$\begin{bmatrix} x \\ y \\ z \end{bmatrix} = \begin{bmatrix} 1.8 & 0.88 & 0.80 \\ 1 & 1.6 & 1 \\ 0.2 & 0.32 & 1.20 \end{bmatrix} \begin{bmatrix} 10 \\ 5 \\ 15 \end{bmatrix} = \begin{bmatrix} 34.4 \\ 33 \\ 21.6 \end{bmatrix}.$$

We conclude that $34.4 million worth of goods of one industry, $33 million worth of a second industry, and $21.6 million worth of a third industry should be produced.

USING TECHNOLOGY EXERCISES 2.7, page 166

1. The final outputs of the first, second, third, and fourth industries are 602.62, 502.30, 572.57, and 523.46 units, respectively.

3. The final outputs of the first, second, third, and fourth industries are 143.06, 132.98, 188.59, and 125.53 units, respectively.

CHAPTER 2, REVIEW EXERCISES, page 168

1. $\begin{bmatrix} 1 & 2 \\ -1 & 3 \\ 2 & 1 \end{bmatrix} + \begin{bmatrix} 1 & 0 \\ 0 & 1 \\ 1 & 2 \end{bmatrix} = \begin{bmatrix} 2 & 2 \\ -1 & 4 \\ 3 & 3 \end{bmatrix}.$ 3. $\begin{bmatrix} -3 & 2 & 1 \end{bmatrix} \begin{bmatrix} 2 & 1 \\ -1 & 0 \\ 2 & 1 \end{bmatrix} = \begin{bmatrix} -6 & -2 \end{bmatrix}$

5. By the equality of matrices, $x = 2$, $z = 1$, $y = 3$ and $w = 3$.

7. By the equality of matrices,

$$a + 3 = 6, \text{ or } a = 3.$$
$$-1 = e + 2, \text{ or } e = -3; b = 4$$
$$c + 1 = -1, \text{ or } c = -2; d = 2$$
$$e + 2 = -1, \text{ and } e = -3.$$

9. $2A + 3B = 2 \begin{bmatrix} 1 & 3 & 1 \\ -2 & 1 & 3 \\ 4 & 0 & 2 \end{bmatrix} + 3 \begin{bmatrix} 2 & 1 & 3 \\ -2 & -1 & -1 \\ 1 & 4 & 2 \end{bmatrix} = \begin{bmatrix} 2 & 6 & 2 \\ -4 & 2 & 6 \\ 8 & 0 & 4 \end{bmatrix} + \begin{bmatrix} 6 & 3 & 9 \\ -6 & -3 & -3 \\ 3 & 12 & 6 \end{bmatrix}$

$$= \begin{bmatrix} 8 & 9 & 11 \\ -10 & -1 & 3 \\ 11 & 12 & 10 \end{bmatrix}.$$

11. $3A = 3 \begin{bmatrix} 1 & 3 & 1 \\ -2 & 1 & 3 \\ 4 & 0 & 2 \end{bmatrix} = \begin{bmatrix} 3 & 9 & 3 \\ -6 & 3 & 9 \\ 12 & 0 & 6 \end{bmatrix}$

and $2(3A) = 2 \begin{bmatrix} 3 & 9 & 3 \\ -6 & 3 & 9 \\ 12 & 0 & 6 \end{bmatrix} = \begin{bmatrix} 6 & 18 & 6 \\ -12 & 6 & 18 \\ 24 & 0 & 12 \end{bmatrix}.$

13.

$$B - C = \begin{bmatrix} 2 & 1 & 3 \\ -2 & -1 & -1 \\ 1 & 4 & 2 \end{bmatrix} - \begin{bmatrix} 3 & -1 & 2 \\ 1 & 6 & 4 \\ 2 & 1 & 3 \end{bmatrix} = \begin{bmatrix} -1 & 2 & 1 \\ -3 & -7 & -5 \\ -1 & 3 & -1 \end{bmatrix}$$

and so $A(B-C) = \begin{bmatrix} 1 & 3 & 1 \\ -2 & 1 & 3 \\ 4 & 0 & 2 \end{bmatrix} \begin{bmatrix} -1 & 2 & 1 \\ -3 & -7 & -5 \\ -1 & 3 & -1 \end{bmatrix} = \begin{bmatrix} -11 & -16 & -15 \\ -4 & -2 & -10 \\ -6 & 14 & 2 \end{bmatrix}$.

15.

$$BC = \begin{bmatrix} 2 & 1 & 3 \\ -2 & -1 & -1 \\ 1 & 4 & 2 \end{bmatrix} \begin{bmatrix} 3 & -1 & 2 \\ 1 & 6 & 4 \\ 2 & 1 & 3 \end{bmatrix} = \begin{bmatrix} 13 & 7 & 17 \\ -9 & -5 & -11 \\ 11 & 25 & 24 \end{bmatrix}$$

$$ABC = \begin{bmatrix} 1 & 3 & 1 \\ -2 & 1 & 3 \\ 4 & 0 & 2 \end{bmatrix} \begin{bmatrix} 13 & 7 & 17 \\ -9 & -5 & -11 \\ 11 & 25 & 24 \end{bmatrix} = \begin{bmatrix} -3 & 17 & 8 \\ -2 & 56 & 27 \\ 74 & 78 & 116 \end{bmatrix}.$$

17. Using the Gauss-Jordan elimination method, we find

$$\begin{bmatrix} 2 & -3 & | & 5 \\ 3 & 4 & | & -1 \end{bmatrix} \xrightarrow{\frac{1}{2}R_1} \begin{bmatrix} 1 & -\frac{3}{2} & | & \frac{5}{2} \\ 3 & 4 & | & -1 \end{bmatrix} \xrightarrow{R_2 - 3R_1} \begin{bmatrix} 1 & -\frac{3}{2} & | & \frac{5}{2} \\ 0 & \frac{17}{2} & | & -\frac{17}{2} \end{bmatrix}$$

$$\xrightarrow{\frac{2}{17}R_2} \begin{bmatrix} 1 & -\frac{3}{2} & | & \frac{5}{2} \\ 0 & 1 & | & -1 \end{bmatrix} \xrightarrow{R_1 + \frac{3}{2}R_2} \begin{bmatrix} 1 & 0 & | & 1 \\ 0 & 1 & | & -1 \end{bmatrix}.$$ We conclude that $x = 1$ and $y = -1$.

19.

$$\begin{bmatrix} 1 & -1 & 2 & | & 5 \\ 3 & 2 & 1 & | & 10 \\ 2 & -3 & -2 & | & -10 \end{bmatrix} \xrightarrow[R_3 - 2R_1]{R_2 - 3R_1} \begin{bmatrix} 1 & -1 & 2 & | & 5 \\ 0 & 5 & -5 & | & -5 \\ 0 & -1 & -6 & | & -20 \end{bmatrix} \xrightarrow{\frac{1}{5}R_2} \begin{bmatrix} 1 & -1 & 2 & | & 5 \\ 0 & 1 & -1 & | & -1 \\ 0 & -1 & -6 & | & -20 \end{bmatrix}$$

$$\xrightarrow[R_3 + R_2]{R_1 + R_2} \begin{bmatrix} 1 & 0 & 1 & | & 4 \\ 0 & 1 & -1 & | & -1 \\ 0 & 0 & -7 & | & -21 \end{bmatrix} \xrightarrow{-\frac{1}{7}R_3} \begin{bmatrix} 1 & 0 & 1 & | & 4 \\ 0 & 1 & -1 & | & -1 \\ 0 & 0 & 1 & | & 3 \end{bmatrix} \xrightarrow[R_2 + R_3]{R_1 - R_3}$$

$$= \begin{bmatrix} 1 & 0 & 0 & | & 1 \\ 0 & 1 & 0 & | & 2 \\ 0 & 0 & 1 & | & 3 \end{bmatrix}. \text{ Therefore, } x = 1, y = 2, \text{ and } z = 3.$$

21. $\begin{bmatrix} 3 & -2 & 4 & | & 11 \\ 2 & -4 & 5 & | & 4 \\ 1 & 2 & -1 & | & 10 \end{bmatrix} \xrightarrow{R_1 - R_2} \begin{bmatrix} 1 & 2 & -1 & | & 7 \\ 2 & -4 & 5 & | & 4 \\ 1 & 2 & -1 & | & 10 \end{bmatrix} \xrightarrow[R_3 - R_1]{R_2 - 2R_1} \begin{bmatrix} 1 & 2 & -1 & | & 7 \\ 0 & -8 & 7 & | & -10 \\ 0 & 0 & 0 & | & 3 \end{bmatrix}.$

Since this last row implies that $0 = 3!$, we conclude that the system has no solution.

23. $\begin{bmatrix} 3 & -2 & 1 & | & 4 \\ 1 & 3 & -4 & | & -3 \\ 2 & -3 & 5 & | & 7 \\ 1 & -8 & 9 & | & 10 \end{bmatrix} \xrightarrow{R_1 - R_3} \begin{bmatrix} 1 & 1 & -4 & | & -3 \\ 1 & 3 & -4 & | & -3 \\ 2 & -3 & 5 & | & 7 \\ 1 & -8 & 9 & | & 10 \end{bmatrix} \xrightarrow[\substack{R_3 - 2R_1 \\ R_4 - R_1}]{R_2 - R_1} \begin{bmatrix} 1 & 1 & -4 & | & -3 \\ 0 & 2 & 0 & | & 0 \\ 0 & -5 & 13 & | & 13 \\ 0 & -9 & 13 & | & 13 \end{bmatrix}$

$\xrightarrow{\frac{1}{2}R_2} \begin{bmatrix} 1 & 1 & -4 & | & -3 \\ 0 & 1 & 0 & | & 0 \\ 0 & -5 & 13 & | & 13 \\ 0 & -9 & 13 & | & 13 \end{bmatrix} \xrightarrow[\substack{R_3 + 5R_2 \\ R_4 + 9R_2}]{R_1 - R_2} \begin{bmatrix} 1 & 0 & -4 & | & -3 \\ 0 & 1 & 0 & | & 0 \\ 0 & 0 & 13 & | & 13 \\ 0 & 0 & 13 & | & 13 \end{bmatrix} \xrightarrow{\frac{1}{13}R_3} \begin{bmatrix} 1 & 0 & -4 & | & -3 \\ 0 & 1 & 0 & | & 0 \\ 0 & 0 & 1 & | & 1 \\ 0 & 0 & 13 & | & 13 \end{bmatrix}$

$\xrightarrow[\substack{R_4 - 13R_3}]{R_1 + 4R_3} \begin{bmatrix} 1 & 0 & 0 & | & 1 \\ 0 & 1 & 0 & | & 0 \\ 0 & 0 & 1 & | & 1 \\ 0 & 0 & 0 & | & 0 \end{bmatrix}.$ Therefore, $x = 1, y = 0,$ and $z = 1.$

25. $A^{-1} = \dfrac{1}{(3)(2) - (1)(1)} \begin{bmatrix} 2 & -1 \\ -1 & 3 \end{bmatrix} = \begin{bmatrix} \frac{2}{5} & -\frac{1}{5} \\ -\frac{1}{5} & \frac{3}{5} \end{bmatrix}.$

27. $A^{-1} = \dfrac{1}{(3)(2) - (2)(4)} \begin{bmatrix} 2 & -4 \\ -2 & 3 \end{bmatrix} = \begin{bmatrix} -1 & 2 \\ 1 & -\frac{3}{2} \end{bmatrix}.$

29. $\begin{bmatrix} 2 & 3 & 1 & | & 1 & 0 & 0 \\ 1 & -1 & 2 & | & 0 & 1 & 0 \\ 1 & 2 & 1 & | & 0 & 0 & 1 \end{bmatrix} \xrightarrow{R_1 - R_2} \begin{bmatrix} 1 & 4 & -1 & | & 1 & -1 & 0 \\ 1 & -1 & 2 & | & 0 & 1 & 0 \\ 1 & 2 & 1 & | & 0 & 0 & 1 \end{bmatrix} \xrightarrow[\substack{R_3 - R_1}]{R_2 - R_1}$

$$\begin{bmatrix} 1 & 4 & -1 & 1 & -1 & 0 \\ 0 & -5 & 3 & -1 & 2 & 0 \\ 0 & -2 & 2 & -1 & 1 & 1 \end{bmatrix} \xrightarrow{R_2-3R_3} \begin{bmatrix} 1 & 4 & -1 & 1 & -1 & 0 \\ 0 & 1 & -3 & 2 & -1 & -3 \\ 0 & -2 & 2 & -1 & 1 & 1 \end{bmatrix} \xrightarrow[R_3+2R_2]{R_1-4R_2}$$

$$\begin{bmatrix} 1 & 0 & 11 & -7 & 3 & 12 \\ 0 & 1 & -3 & 2 & -1 & -3 \\ 0 & 0 & -4 & 3 & -1 & -5 \end{bmatrix} \xrightarrow{-\frac{1}{4}R_3} \begin{bmatrix} 1 & 0 & 11 & -7 & 3 & 12 \\ 0 & 1 & -3 & 2 & -1 & -3 \\ 0 & 0 & 1 & -\frac{3}{4} & \frac{1}{4} & \frac{5}{4} \end{bmatrix} \xrightarrow[R_2+3R_3]{R_1-11R_3}$$

$$\begin{bmatrix} 1 & 0 & 0 & \frac{5}{4} & \frac{1}{4} & -\frac{7}{4} \\ 0 & 1 & 0 & -\frac{1}{4} & -\frac{1}{4} & \frac{3}{4} \\ 0 & 0 & 1 & -\frac{3}{4} & \frac{1}{4} & \frac{5}{4} \end{bmatrix}. \text{ So } A^{-1} = \begin{bmatrix} \frac{5}{4} & \frac{1}{4} & -\frac{7}{4} \\ -\frac{1}{4} & -\frac{1}{4} & \frac{3}{4} \\ -\frac{3}{4} & \frac{1}{4} & \frac{5}{4} \end{bmatrix}.$$

31. $\begin{bmatrix} 1 & 2 & 4 & 1 & 0 & 0 \\ 3 & 1 & 2 & 0 & 1 & 0 \\ 1 & 0 & -6 & 0 & 0 & 1 \end{bmatrix} \xrightarrow[R_3-R_1]{R_2-3R_1} \begin{bmatrix} 1 & 2 & 4 & 1 & 0 & 0 \\ 0 & -5 & -10 & -3 & 1 & 0 \\ 0 & -2 & -10 & -1 & 0 & 1 \end{bmatrix} \xrightarrow{R_2-3R_3}$

$$\begin{bmatrix} 1 & 2 & 4 & 1 & 0 & 0 \\ 0 & 1 & 20 & 0 & 1 & -3 \\ 0 & -2 & -10 & -1 & 0 & 1 \end{bmatrix} \xrightarrow[R_3+2R_2]{R_1-2R_2} \begin{bmatrix} 1 & 0 & -36 & 1 & -2 & 6 \\ 0 & 1 & 20 & 0 & 1 & -3 \\ 0 & 0 & 30 & -1 & 2 & -5 \end{bmatrix} \xrightarrow{\frac{1}{30}R_3}$$

$$\begin{bmatrix} 1 & 0 & -36 & 1 & -2 & 6 \\ 0 & 1 & 20 & 0 & 1 & -3 \\ 0 & 0 & 1 & -\frac{1}{30} & \frac{1}{15} & -\frac{1}{6} \end{bmatrix} \xrightarrow[R_2-20R_3]{R_1+36R_3} \begin{bmatrix} 1 & 0 & 0 & -\frac{1}{5} & \frac{2}{5} & 0 \\ 0 & 1 & 0 & \frac{2}{3} & -\frac{1}{3} & \frac{1}{3} \\ 0 & 0 & 1 & -\frac{1}{30} & \frac{1}{15} & -\frac{1}{6} \end{bmatrix}$$

So $A^{-1} = \begin{bmatrix} -\frac{1}{5} & \frac{2}{5} & 0 \\ \frac{2}{3} & -\frac{1}{3} & \frac{1}{3} \\ -\frac{1}{30} & \frac{1}{15} & -\frac{1}{6} \end{bmatrix}.$

33. $(A^{-1}B)^{-1} = B^{-1}(A^{-1})^{-1} = B^{-1}A$. Now

$$B^{-1} = \frac{1}{(3)(2)-4(1)} \begin{bmatrix} 2 & -1 \\ -4 & 3 \end{bmatrix} = \begin{bmatrix} 1 & -\frac{1}{2} \\ -2 & \frac{3}{2} \end{bmatrix}. \quad B^{-1}A = \begin{bmatrix} 1 & -\frac{1}{2} \\ -2 & \frac{3}{2} \end{bmatrix}\begin{bmatrix} 1 & 2 \\ -1 & 2 \end{bmatrix} = \begin{bmatrix} \frac{3}{2} & 1 \\ -\frac{7}{2} & -1 \end{bmatrix}.$$

35. $\quad 2A - C = \begin{bmatrix} 2 & 4 \\ -2 & 4 \end{bmatrix} - \begin{bmatrix} 1 & 1 \\ -1 & 2 \end{bmatrix} = \begin{bmatrix} 1 & 3 \\ -1 & 2 \end{bmatrix}.$

$$(2A - C)^{-1} = \frac{1}{(1)(2) - (-1)(3)} \begin{bmatrix} 2 & -3 \\ 1 & 1 \end{bmatrix} = \begin{bmatrix} \frac{2}{5} & -\frac{3}{5} \\ \frac{1}{5} & \frac{1}{5} \end{bmatrix}.$$

37. $\quad A = \begin{bmatrix} 2 & 3 \\ 1 & -2 \end{bmatrix}, \quad X = \begin{bmatrix} x \\ y \end{bmatrix}, \quad C = \begin{bmatrix} -8 \\ 3 \end{bmatrix}; \quad A^{-1} = \frac{1}{(-2)(2) - (1)(3)} \begin{bmatrix} -2 & -3 \\ -1 & 2 \end{bmatrix} = \begin{bmatrix} \frac{2}{7} & \frac{3}{7} \\ \frac{1}{7} & -\frac{2}{7} \end{bmatrix}$

$$\begin{bmatrix} x \\ y \end{bmatrix} = A^{-1}B = \begin{bmatrix} \frac{2}{7} & \frac{3}{7} \\ \frac{1}{7} & -\frac{2}{7} \end{bmatrix} \begin{bmatrix} -8 \\ 3 \end{bmatrix} = \begin{bmatrix} -1 \\ -2 \end{bmatrix}.$$

39. Put

$$X = \begin{bmatrix} x \\ y \\ z \end{bmatrix}, \quad A = \begin{bmatrix} 1 & -2 & 4 \\ 2 & 3 & -2 \\ 1 & 4 & -6 \end{bmatrix}, \quad C = \begin{bmatrix} 13 \\ 0 \\ -15 \end{bmatrix}.$$

Then $AX = C$ and $X = A^{-1}C$. To find A^{-1},

$$\begin{bmatrix} 1 & -2 & 4 & | & 1 & 0 & 0 \\ 2 & 3 & -2 & | & 0 & 1 & 0 \\ 1 & 4 & -6 & | & 0 & 0 & 1 \end{bmatrix} \xrightarrow[R_3 - R_1]{R_2 - 2R_1} \begin{bmatrix} 1 & -2 & 4 & | & 1 & 0 & 0 \\ 0 & 7 & -10 & | & -2 & 1 & 0 \\ 0 & 6 & -10 & | & -1 & 0 & 1 \end{bmatrix} \xrightarrow{R_2 - R_3}$$

$$\begin{bmatrix} 1 & -2 & 4 & | & 1 & 0 & 0 \\ 0 & 1 & 0 & | & -1 & 1 & -1 \\ 0 & 6 & -10 & | & -1 & 0 & 1 \end{bmatrix} \xrightarrow[R_3 - 6R_2]{R_1 + 2R_2} \begin{bmatrix} 1 & 0 & 4 & | & -1 & 2 & -2 \\ 0 & 1 & 0 & | & -1 & 1 & -1 \\ 0 & 0 & -10 & | & 5 & -6 & 7 \end{bmatrix} \xrightarrow{-\frac{1}{10}R_3}$$

$$\begin{bmatrix} 1 & 0 & 4 & | & -1 & 2 & -2 \\ 0 & 1 & 0 & | & -1 & 1 & -1 \\ 0 & 0 & 1 & | & -\frac{1}{2} & \frac{3}{5} & -\frac{7}{10} \end{bmatrix} \xrightarrow{R_1 - 4R_3} \begin{bmatrix} 1 & 0 & 0 & | & 1 & -\frac{2}{5} & \frac{4}{5} \\ 0 & 1 & 0 & | & -1 & 1 & -1 \\ 0 & 0 & 1 & | & -\frac{1}{2} & \frac{3}{5} & -\frac{7}{10} \end{bmatrix}.$$

So $A^{-1} = \begin{bmatrix} 1 & -\frac{2}{5} & \frac{4}{5} \\ -1 & 1 & -1 \\ -\frac{1}{2} & \frac{3}{5} & -\frac{7}{10} \end{bmatrix}$. Therefore, $X = A^{-1}C = \begin{bmatrix} 1 & -\frac{2}{5} & \frac{4}{5} \\ -1 & 1 & -1 \\ -\frac{1}{2} & \frac{3}{5} & -\frac{7}{10} \end{bmatrix} \begin{bmatrix} 13 \\ 0 \\ -15 \end{bmatrix} = \begin{bmatrix} 1 \\ 2 \\ 4 \end{bmatrix}$

that is, $x = 1$, $y = 2$, and $z = 4$.

41.   Let

$$S = \begin{array}{c} \\ \\ \\ \\ \end{array} \begin{matrix} \text{Prem} & \text{Sup} & \text{Reg} & \text{Dies} \\ \begin{bmatrix} 600 & 800 & 1000 & 700 \\ 700 & 600 & 1200 & 400 \\ 900 & 700 & 1400 & 800 \end{bmatrix} \end{matrix} \quad \text{and} \quad T = \begin{matrix} \text{Prem} \\ \text{Sup} \\ \text{Reg} \\ \text{Dies} \end{matrix} \begin{bmatrix} 1.60 \\ 1.40 \\ 1.20 \\ 1.50 \end{bmatrix}$$

be the matrices representing the sales in the three gasoline stations and the unit prices for the various fuels, respectively. Then the total revenue at each station is found by computing

$$ST = \begin{bmatrix} 600 & 800 & 1000 & 700 \\ 700 & 600 & 1200 & 400 \\ 900 & 700 & 1400 & 800 \end{bmatrix} \begin{bmatrix} 1.60 \\ 1.40 \\ 1.20 \\ 1.50 \end{bmatrix} = \begin{bmatrix} 4330 \\ 4000 \\ 5300 \end{bmatrix}.$$

We conclude that the total revenue of station A is $4330, that of station B is $4000, and that of station C is $5300.

43.   We wish to solve the system of equations
$$\begin{aligned} 2x + 2y + 3z &= 210 \\ 2x + 3y + 4z &= 270 \\ 3x + 4y + 3z &= 300. \end{aligned}$$
Using the Gauss–Jordan method of elimination, we find

$$\begin{bmatrix} 2 & 2 & 3 & 210 \\ 2 & 3 & 4 & 270 \\ 3 & 4 & 3 & 300 \end{bmatrix} \xrightarrow{\frac{1}{2}R_1} \begin{bmatrix} 1 & 1 & \frac{3}{2} & 105 \\ 2 & 3 & 4 & 270 \\ 3 & 4 & 3 & 300 \end{bmatrix} \xrightarrow[R_3 - 3R_1]{R_2 - 2R_1} \begin{bmatrix} 1 & 1 & \frac{3}{2} & 105 \\ 0 & 1 & 1 & 60 \\ 0 & 1 & -\frac{3}{2} & -15 \end{bmatrix}$$

$$\xrightarrow[R_3 - R_2]{R_1 - R_2} \begin{bmatrix} 1 & 0 & \frac{1}{2} & 45 \\ 0 & 1 & 1 & 60 \\ 0 & 0 & -\frac{5}{2} & -75 \end{bmatrix} \xrightarrow{-\frac{2}{5}R_3} \begin{bmatrix} 1 & 0 & \frac{1}{2} & 45 \\ 0 & 1 & 1 & 60 \\ 0 & 0 & 1 & 30 \end{bmatrix} \xrightarrow[R_2 - R_3]{R_1 - \frac{1}{2}R_3} \begin{bmatrix} 1 & 0 & 0 & 30 \\ 0 & 1 & 0 & 30 \\ 0 & 0 & 1 & 30 \end{bmatrix}$$

So $x = y = z = 30$. Therefore, Desmond should produce 30 of each type of pendant.

CHAPTER 3

EXERCISES 3.1, page 176

1. $4x - 8 < 0$ implies $x < 2$. The graph of the inequality follows.

3. $x - y \leq 0$ implies $x \leq y$. The graph of the inequality follows.

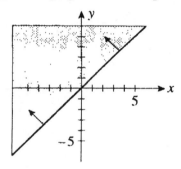

5. The graph of the inequality $x \leq -3$ follows.

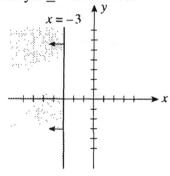

7. We first sketch the straight line with equation $2x + y = 4$. Next, picking the test point $(0,0)$, we have $2(0) + (0) = 0 \leq 4$. We conclude that the half-plane containing the origin is the required half-plane.

9. We first sketch the graph of the straight line $4x - 3y = -24$. Next, picking the test point $(0,0)$, we see that $4(0) - 3(0) = 0 \not< -24$. We conclude that the half-plane not containing the origin is the required half-plane. The graph of this inequality follows.

11. The system of linear inequalities that describes the shaded region is
    $$x \geq 1, x \leq 5, y \geq 2, \text{ and } y \leq 4.$$
    We may also combine the first and second inequalities and the third and fourth inequalities and write
    $$1 \leq x \leq 5 \quad \text{and} \quad 2 \leq y \leq 4.$$

13. The system of linear inequalities that describes the shaded region is
    $$2x - y \geq 2, 5x + 7y \geq 35, \text{ and } x \leq 4.$$

15. The system of linear inequalities that describes the shaded region is
    $$7x + 4y \leq 140, x + 3y \geq 30, \text{ and } x - y \geq -10.$$

17. The system of linear inequalities that describes the shaded region is
    $$x + y \geq 7, x \geq 2, y \geq 3, \text{ and } y \leq 7.$$

19. The required solution set is shown below.

To find the coordinates of A, we solve the system
$$2x + 4y = 16$$
$$-x + 3y = 7,$$

giving $A = (2,3)$. Observe that a dotted line is used to show that no point on the line constitutes a solution to the given problem. Observe that this is an unbounded solution set.

21. The solution set is shown in the figure below. Observe that the set is unbounded.

To find the coordinates of A, we solve the system $\begin{cases} x - y = 0 \\ 2x + 3y = 10 \end{cases}$ giving $A = (2,2)$.

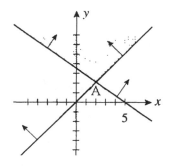

23. The half-planes defined by the two inequalities are shown in the figure at the right. Since the two half-planes have no points in common, we conclude that the given system of inequalities has no solution. (The empty set is a bounded set.)

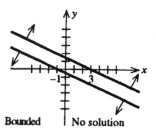

Bounded   No solution

25. The half-planes defined by the three inequalities are shown below. The point A is

found by solving the system $\begin{cases} x+y=6 \\ x \quad =3 \end{cases}$ giving $A = (3,3)$. Observe that this is a bounded solution set.

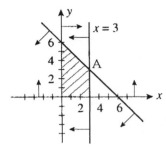

27. The half-planes defined by the given inequalities are shown below. Observe that the two lines described by the equations $3x - 6y = 12$ and $-x + 2y = 4$ do not intersect because they are parallel. The solution set is unbounded.

29. The required solution set is shown in the figure below. The coordinates of A are found by solving the system $\begin{array}{c} 3x - 7y = -24 \\ x + 3y = 8 \end{array}$ giving $(-1,3)$. The solution set is unbounded.

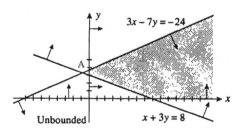

31. The required solution set is shown in the figure that follows. The solution set is bounded.

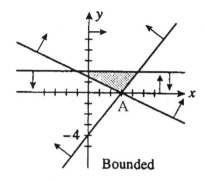

Bounded

33. The required solution set is shown in the figure below. The solution set has vertices at $(0,6)$, $(5,0)$, $(4,0)$, and $(1,3)$. The solution set is bounded.

$x + y = 4$

$6x + 5y = 30$

Bounded

$3x + y = 6$

35. The required solution set is shown in the figure below. The unbounded solution set has vertices at $(2,8)$, $(0,6)$, $(0,3)$, and $(2,2)$.

$x - y = -6$

$x - 2y = -2$

$x - 2y = -14$

37. False. It is always a half-plane. A straight line is the graph of a linear equation and vice-versa.

39. True. Since a circle can always be enclosed by a rectangle, the solution set of such a

system is bounded if it can be enclosed by a rectangle.

EXERCISES 3.2, page 185

1. We tabulate the given information:

	Product A	Product B	Time Available
Machine I	6	9	300
Machine II	5	4	180
Profit per unit ($)	3	4	

    Let x and y denote the number of units of Product A and Product B to be produced. Then the required linear programming problem is:

    Maximize $P = 3x + 4y$ subject to the constraints
    $$6x + 9y \le 300$$
    $$5x + 4y \le 180$$
    $$x \ge 0, y \ge 0$$

3. Let x denote the number of model A grates to be produced and y denote the number of model B grates to be produced. Since only 1000 pounds of cast iron are available, we must have
    $$3x + 4y \le 1000.$$
    The restriction that only 20 hours of labor are available per day implies that
    $$6x + 3y \le 1200. \quad \text{(time in minutes)}$$

    Then the profit on the production of these grates is given by
    $$P = 2x + 1.5y.$$
    The additional restriction that at least 150 model A grates be produced each day implies that $x \ge 150$. Summarizing, we have the following linear programming problem:

    Maximize $P = 2x + 1.5y$ subject to
    $$3x + 4y \le 1000$$
    $$6x + 3y \le 1200$$
    $$x \ge 150, y \ge 0$$

5. Let x and y denote the amount (in thousands of dollars) to be invested in project A

and project B, respectively. Since the amount available for investment is up to $500,000, we have $x + y \leq 500$. Next, the condition on the allocation of the funds implies that

$$y \leq 0.4(x + y), \quad -0.4x + 0.6y \leq 0, \quad \text{or} \quad -2x + 3y \leq 0$$

The linear programming problem at hand is

Maximize $P = 0.1x + 0.15y$ subject to

$$x + y \leq 500$$
$$-2x + 3y \leq 0$$
$$x \geq 0, y \geq 0$$

7. Let x denote the number of fully assembled units to be produced daily and let y denote the number of kits to be produced. Then the fraction of the day the fabrication department works on the fully assembled cabinets is $\frac{1}{200}x$. Similarly the fraction of the day the fabrication department works on kits is $\frac{1}{200}y$. Since the fraction of the day during which the fabrication department is busy cannot exceed one, we must have

$$\frac{1}{200}x + \frac{1}{200}y \leq 1.$$

Similarly, the restrictions place on the assembly department leads to the inequality

$$\frac{1}{100}x + \frac{1}{300}y \leq 1.$$

The profit (objective) function is $P = 50x + 40y$. Summarizing, the required linear programming problem is

Maximize $P = 50x + 40y$ subject to

$$\frac{1}{200}x + \frac{1}{200}y \leq 1$$
$$\frac{1}{100}x + \frac{1}{300}y \leq 1$$
$$x \geq 0, \ y \geq 0.$$

9. Let x and y denote the number of days the Saddle Mine and the Horseshoe Mine are operated, respectively. Then the operating cost is $C = 14{,}000x + 16{,}000y$. The amount of gold produced in the two mines is $(50x + 75y)$ oz, and this amount must be at least 650 oz. So we have $50x + 75y \geq 650$. Similarly, the requirement for silver production leads to the inequality $3000x + 1000y \geq 18{,}000$. So the problem is

$$\text{Minimize } C = 14{,}000x + 16{,}000y \text{ subject to}$$
$$50x + \quad 75y \geq 650$$
$$3000x + 1000y \geq 18{,}000$$
$$x \geq 0, \; y \geq 0$$

11. Let x, y, and z denote the amount of money she invests in project A, project B, and project C, respectively. Since she plans to invest up to \$2 million, we must have $x + y + z \leq 2{,}000{,}000$. Because she decides to put not more than 20% of her total investment in project C, we have
$$z \leq 0.2(x + y + z) \quad \text{or} \quad -0.2x - 0.2y + 0.8z \leq 0$$
Since her investments in project B and C should not exceed 60% of her total investment, we have
$$y + z \leq 0.6(x + y + z) \quad \text{or} \quad -0.6x + 0.4y + 0.4z \leq 0$$
Also, since her investment in project A should be at least 60% of her investments in

projects B and C, we have
$$x \geq 0.6(y + z) \quad \text{or} \quad -x + 0.6y + 0.6z \leq 0$$
Finally, the returns on her investments are given by $P = 0.1x + 0.15y + 0.2z$.
To summarize, the problem is
$$\text{Maximize } P = 0.1x + 0.15y + 0.2z \text{ subject to}$$
$$x + \quad y + \quad z \leq 2{,}000{,}000$$
$$-2x - 2y + 8z \leq 0$$
$$-6x + 4y + 4z \leq 0$$
$$-10x + 6y + 6z \leq 0$$
$$x \geq 0, \, y \geq 0, \, z \geq 0$$

13. Let x and y denote the amount of money given in aid to country A and country B, respectively. Then the requirement that country A receives between \$1 and \$1.5 million, inclusive, in aid, implies $1 \leq x \leq 1.5$. Similarly, the requirement that country B receive at least \$0.75 million in aid implies $y \geq 0.75$. The condition that between \$2 and \$2.5 million dollars in aid has been earmarked for these two countries implies that
$$x + y \leq 2.5$$
$$x + y \geq 2$$
Therefore, we have the following linear programming problem:

Maximize $P = 0.6x + 0.8y$ subject to
$$x + y \le 2.5$$
$$x + y \ge 2$$
$$1 \le x \le 1.5$$
$$y \ge 0.75$$

15. Let x denote the number of picture tubes shipped from location I to city A and let y denote the number of picture tubes shipped from location I to city B. Since the number of picture tubes required by the two factories in city A and city B are 3000 and 4000, respectively, the number of picture tubes shipped from location II to city A and city B, are (3000 - x) and (4000 - y), respectively. These numbers are shown in the following schematic,

Referring to the schematic and the shipping schedule, we find that the total shipping costs incurred by the company are given by
$$C = 3x + 2y + 4(3000 - x) + 5(4000 - y)$$
$$= 32000 - x - 3y$$
The production constraints on Location I and II lead to the inequalities
$$x + \qquad y \le 6000$$
$$(3000 - x) + (4000 - y) \le 5000$$
This last inequality simplifies to $x + y \ge 2000$.
The requirements of the two factories lead to the inequalities
$$x \ge 0,\, y \ge 0,\, 3000 - x \ge 0,\text{ and } 4000 - y \ge 0.$$
These last two inequalities may be written as $x \le 3000$ and $y \le 4000$.
Summarizing, we have the following linear programming problem:
Minimize $C = 32{,}000 - x - 3y$ subject to
$$x + y \le 6000$$
$$x + y \ge 2000$$
$$x \le 3000$$
$$y \le 4000$$
$$x \ge 0,\ y \ge 0$$

17. Let x, y, and z denote the number of units produced of products A, B, and C, respectively. From the given information, we formulate the following linear programming problem:

Maximize $P = 18x + 12y + 15z$ subject to

$$2x + y + 2z \le 900$$
$$3x + y + 2z \le 1080$$
$$2x + 2y + z \le 840$$
$$x \ge 0, y \ge 0, z \ge 0$$

19. We first tabulate the given information:

Department	Model A	Model B	Model C	Time Available
Fabrication	$\frac{5}{4}$	$\frac{3}{2}$	$\frac{3}{2}$	310
Assembly	1	1	$\frac{3}{4}$	205
Finishing	1	1	$\frac{1}{2}$	190

Let x, y, and z denote the number of units of model A, model B, and model C to be produced, respectively. Then the required linear programming problem is

Maximize $P = 26x + 28y + 24z$ subject to

$$\tfrac{5}{4}x + \tfrac{3}{2}y + \tfrac{3}{2}z \le 310$$
$$x + y + \tfrac{3}{4}z \le 205$$
$$x + y + \tfrac{1}{2}z \le 190$$
$$x \ge 0, y \ge 0, z \ge 0$$

21. The shipping costs are tabulated in the following table.

	Warehouse A	Warehouse B	Warehouse C
Plant I	60	60	80
Plant II	80	70	50

Letting x_1 denote the number of pianos shipped from plant I to warehouse A, x_2 the number of pianos shipped from plant I to warehouse B, and so we have

	Warehouse A	Warehouse B	Warehouse C	Maximum Production
Plant I	x_1	x_2	x_3	300
Plant II	x_4	x_5	x_6	250
Minimum Requirement	200	150	200	

From the two tables we see that the total monthly shipping cost is given by
$$C = 60x_1 + 60x_2 + 80x_3 + 80x_4 + 70x_5 + 50x_6.$$
Next, the production constraints on plants I and II lead to the inequalities
$$x_1 + x_4 \geq 200$$
$$x_2 + x_5 \geq 150$$
$$x_3 + x_6 \geq 200$$
Summarizing we have the following linear programming problem:
Minimize $C = 60x_1 + 60x_2 + 80x_3 + 80x_4 + 70x_5 + 50x_6$ subject to
$$x_1 + x_2 + x_3 \leq 300$$
$$x_4 + x_5 + x_6 \leq 250$$
$$x_1 + x_4 \geq 200$$
$$x_2 + x_5 \geq 150$$
$$x_3 + x_6 \geq 200$$
$$x_1 \geq 0, \ x_2 \geq 0, \ ..., \ x_6 \geq 0$$

23. The given data can be summarized as follows:

	Concentrates			Profit ($)
	Pineapple	Orange	Banana	
Pineapple-orange	8	8	0	1
Orange-banana	0	12	4	0.80
Pineapple-orange-banana	4	8	4	0.90
Maximum available (oz)	16,000	24,000	5,000	

Suppose x, y, and z cartons of pineapple-orange, orange-banana, and pineapple-orange-banana juice are to be produced, respectively. The linear programming problem is
Maximize $P = x + 0.8y + 0.9z$ subject to

$$8x + \qquad 4z \leq 16,000$$
$$8x + 12y + 8z \leq 24,000$$
$$4y + \qquad 4z \leq 5,000$$
$$z \leq 800$$
$$x \geq 0, y \geq 0, \ z \leq 0$$

25. True. It satisfies the definition of a linear programming problem.

EXERCISES 3.3, page 196

1. Evaluating the objective function at each of the corner points we obtain the following table.

Vertex	$Z = 2x + 3y$
(1,1)	$\boxed{5}$
(8,5)	31
(4,9)	$\boxed{35}$
(2,8)	28

From the table, we conclude that the maximum value of Z is 35 and it occurs at the vertex (4,9). The minimum value of Z is 5 and it occurs at the vertex (1,1).

3. Evaluating the objective function at each of the corner points we obtain the following table.

Vertex	$Z = 3x + 4y$
(0,20)	80
(3,10)	49
(4,6)	36
(9,0)	$\boxed{27}$

From the graph, we conclude that there is no maximum value since Z is unbounded. The minimum value of Z is 27 and it occurs at the vertex (9,0).

5. Evaluating the objective function at each of the corner points we obtain the following table.

Vertex	$Z = x + 4y$
(0,6)	24
(4,10)	44
(12,8)	44
(15,0)	15

From the table, we conclude that the maximum value of Z is 44 and it occurs at every point on the line segment joining the points (4,10) and (12,8). The minimum value of Z is 15 and it occurs at the vertex (15,0).

7. The problem is to maximize $P = 2x + 3y$ subject to

$$x + y \le 6$$

$$x \le 3$$

$$x \ge 0, y \ge 0$$

The feasible set S for the problem is shown in the following figure, and the values of the function P at the vertices of S are summarized in the accompanying table.

Vertex	$P = 2x + 3y$
A(0,0)	0
B(3,0)	6
C(3,3)	15
D(0,6)	18

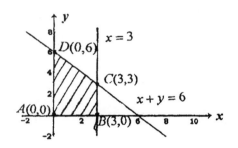

We conclude that P attains a maximum value of 18 when $x = 0$ and $y = 6$.

9. The problem is to maximize $P = 2x + y$ subject to

$$x + y \le 4$$

$$2x + y \le 5$$

$$x \ge 0, y \ge 0$$

Referring to the following figure and table

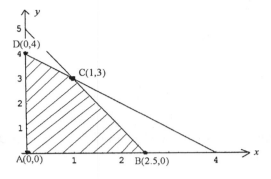

V	$P = 2x + y$
(0,0)	0
(2.5,0)	$\boxed{5}$
(1,3)	$\boxed{5}$
(0,4)	4

we conclude that P attains a maximum value of 5 at any point (x, y) lying on the line segment joining $(1, 3)$ to $(2.5, 0)$.

11. The problem is

$$\text{Maximize } P = x + 8y \text{ subject to}$$
$$x + y \leq 8$$
$$2x + y \leq 10$$
$$x \geq 0, y \geq 0$$

From the following figure and table

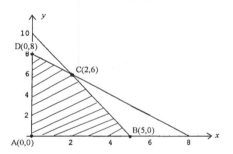

V	$P = x + 8y$
A(0,0)	0
B(5,0)	5
C(2,6)	50
D(0,8)	$\boxed{64}$

We conclude that P attains a maximum value of 64 when $x = 0$ and $y = 8$.

13. The linear programming problem is

$$\text{Maximize } P = x + 3y \text{ subject to}$$
$$2x + y \leq 6$$
$$x + y \leq 4$$
$$x \leq 1$$
$$x \geq 0, y \geq 0$$

From the following figure and table,

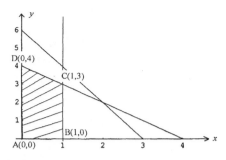

V	$P = x + 3y$
A(0,0)	0
B(1,0)	1
C(1,3)	10
D(0,4)	12

We conclude that P attains a maximum value of 12 when $x = 0$ and $y = 4$.

15. The linear programming problem is

Minimize $C = 3x + 4y$ subject to

$x + \ y \geq 3$

$x + 2y \geq 4$

$x \geq 0, y \geq 0$

From the following figure and table,

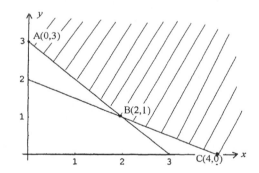

V	$C = 3x + 4y$
A(0,3)	12
B(2,1)	$\boxed{10}$
C(4,0)	12

We conclude that C attains a minimum value of 10 when $x = 2$ and $y = 1$.

17. The linear programming problem is

Minimize $C = 3x + 6y$ subject to

$x + 2y \geq 40$

$x + \ y \geq 30$

$x \geq 0, y \geq 0$

From the following figure and table,

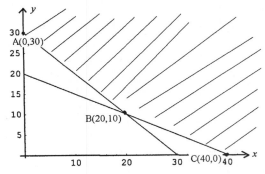

we conclude that C attains a minimum value of 120 at any point on the line segment joining (20,10) to (40,0).

19. The problem is
Minimize $C = 2x + 10y$ subject to
$$5x + 2y \geq 40$$
$$x + 2y \geq 20$$
$$y \geq 3, \ x \geq 0$$

The feasible set S for the problem is shown in the following figure and the values of the function C at the vertices of S are summarized in the accompanying table.

Vertex	$C = 2x + 10y$
A(0,20)	200
B(5, $\frac{15}{2}$)	85
C(14,3)	58

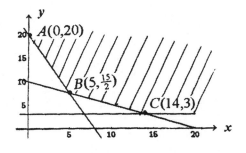

We conclude that C attains a minimum value of 58 when $x = 14$ and $y = 3$.

21. The problem is to minimize $C = 10x + 15y$ subject to
$$x + y \leq 10$$
$$3x + y \geq 12$$
$$-2x + 3y \geq 3$$
$$x \geq 0, \ y \geq 0$$

The feasible set is shown in the following figure, and the values of C at each of the vertices of S are shown in the accompanying table.

Vertex	$C = 10x + 15y$
A(3,3)	75
B($\frac{27}{5}$, $\frac{23}{5}$)	123
C(1,9)	145

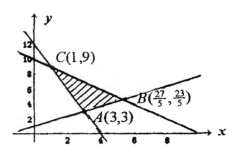

We conclude that C attains a minimum value of 75 when $x = 3$ and $y = 3$.

23. The problem is to maximize $P = 3x + 4y$ subject to
$$x + 2y \le 50$$
$$5x + 4y \le 145$$
$$2x + y \ge 25$$
$$y \ge 5, x \ge 0$$

The feasible set S is shown in the figure that follows, and the values of P at each of the vertices of S are shown in the accompanying table.

Vertex	$P = 3x + 4y$
A(10,5)	50
B(25,5)	95
C(15,$\frac{35}{2}$)	115
D(0,25)	100

We conclude that P attains a maximum value of 115 when $x = 15$ and $y = 35/2$.

25. The problem is to maximize $P = 2x + 3y$ subject to
$$x + y \le 48$$
$$x + 3y \ge 60$$
$$9x + 5y \le 320$$
$$x \ge 10, y \ge 0$$

The feasible set S is shown in the figure that follows, and the values of P at each of the vertices of S are shown in the accompanying table.

Vertex	$P = 2x + 3y$
A(10, $\frac{50}{3}$)	70
B(30,10)	90
C(20,28)	124
D(10,38)	134

We conclude that P attains a maximum value of 134 when $x = 10$ and $y = 38$.

27.   The problem is to find the maximum and minimum value of $P = 10x + 12y$ subject to

$$5x + 2y \geq 63$$
$$x + y \geq 18$$
$$3x + 2y \leq 51$$
$$x \geq 0, y \geq 0$$

The feasible set is shown at the right and the value of P at each of the vertices of S are shown in the accompanying table.

Vertex	$P = 10x + 12y$
A(9,9)	198
B(15,3)	186
C(6, $\frac{33}{2}$)	258

P attains a maximum value of 258 when $x = 6$ and $y = 33/2$. The minimum value of P is 186. It is attained when $x = 15$ and $y = 3$.

29.   The problem is to find the maximum and minimum value of $P = 2x + 4y$ subject to

$$x + y \leq 20$$
$$-x + y \leq 10$$
$$x \leq 10$$
$$x + y \geq 5$$
$$y \geq 5, \ x \geq 0$$

The feasible set is shown in the figure that follows, and the value of P at each of the vertices of S are shown in the accompanying table.

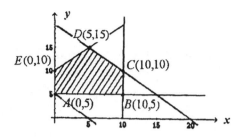

Vertex	$P = 2x + 4y$
A(0,5)	20
B(10,5)	40
C(10,10)	60
D(5,15)	70
E(0,10)	40

P attains a maximum value of 70 when $x = 5$ and $y = 15$. The minimum value of P is 20. It is attained when $x = 0$ and $y = 5$.

31.   Let x and y denote the number of model A and model B fax machines produced in each shift. Then the restriction on manufacturing costs implies
$$100x + 150y \leq 600,000,$$
and the limitation on the number produced implies
$$x + y \leq 2,500.$$
The total profit is $P = 25x + 40y$. Summarizing, we have the following linear programming problem.

$$\text{Maximize } P = 25x + 40y \text{ subject to}$$
$$100x + 150y \leq 600,000$$
$$x + \quad y \leq \quad 2,500$$
$$x \geq 0, \ y \geq 0$$

The graph of the feasible set S and the associated table of values of P follow.

Vertex	$C = 25x + 40y$
A(0,0)	0
B(2500,0)	62,500
C(1500,1000)	77,500
D(0,2500)	100,000

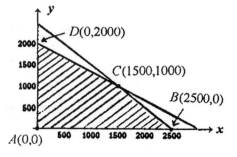

P attains a maximum value of 100,000 when $x = 0$ and $y = 2500$. Thus, by producing 2500 model B fax machines in each shift, the company will realize an optimal profit of $100,000.

33. Refer to the solution of Exercise 2, Section 3.2, The problem is

$$\text{Maximize } P = 2x + 1.5y \text{ subject to}$$
$$3x + 4y \le 1000$$
$$6x + 3y \le 1200$$
$$x \ge 0, \ y \ge 0$$

The graph of the feasible set S and the associated table of values of P follow.

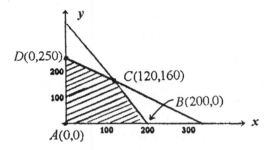

Vertex	$P = 2x + 1.5y$
A(0,0)	0
B(200,0)	400
C(120,160)	480
D(0,250)	375

P attains a maximum value of 480 when $x = 120$ and $y = 160$. Thus, by producing 120 model A grates and 160 model B grates in each shift, the company will realize an optimal profit of $480.

35. Refer to the solution of Exercise 4, Section 3.2. The linear programming problem is

$$\text{Maximize } P = 0.1x + 0.12y \text{ subject to}$$
$$x + y \le 20$$
$$x - 4y \ge 0$$
$$x \ge 0, \ y \ge 0$$

The feasible set S for the problem is shown in the figure at the right, and the value of P at each of the vertices of S is shown in the accompanying table.

Vertex	$C = 0.1x + 0.12y$
A(0,0)	0
B(16,4)	2.08
C(20,0)	2.00

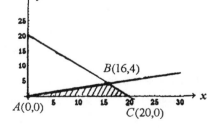

The maximum value of P is attained when $x = 16$ and $y = 4$. Thus, by extending $16 million in housing loans and $4 million in automobile loans, the company will realize a return of $2.08 million on its loans.

37. Refer to Exercise 8, Section 3.2. The problem is

$$\text{Maximize } P = 150x + 200y \text{ subject to}$$
$$40x + 60y \le 7400$$
$$20x + 25y \le 3300$$
$$x \ge 0, \ y \ge 0$$

The graph of the feasible set S and the associated table of values of P follow.

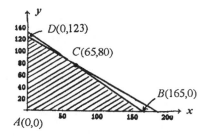

Vertex	$P = 150x + 200y$
A(0,0)	0
B(165,0)	24,750
C(65,80)	25,750
D(0,123)	24,600

P attains a maximum value of 25,750 when $x = 65$ and $y = 80$. Thus, by producing 65 acres of crop A and 80 acres of crop B, the farmer will realize a maximum profit of $25,750.

39. Let x and y denote the number of days the Saddle Mine and the Horseshoe Mine are operated, respectively. Then the operating cost is $C = 14{,}000x + 16{,}000y$. The amount of gold produced in the two mines is $(50x + 75y)$ oz, and this amount must be at least 650 oz. So we have $50x + 75y \ge 650$. Similarly, the requirement for silver production leads to the inequality $3000x + 1000y \ge 18{,}000$. So the problem is

$$\text{Minimize } C = 14{,}000x + 16{,}000y \text{ subject to}$$
$$50x + \ \ 75y \ge 650$$
$$3000x + 1000y \ge 18{,}000$$
$$x \ge 0, \ y \ge 0$$

The feasible set is shown in the accompanying figure. From the table, we see that the minimum value of $C = 152{,}000$ is attained at $x = 4$ and $y = 6$. So, the Saddle Mine should be operated for 4 days and the Horseshoe Mine should be operated for 6 days at a minimum cost of $152,000/day.

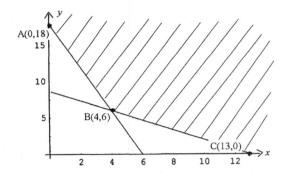

Vertex	$C = 14{,}000x + 16{,}000y$
A(0, 18)	288,000
B(4, 6)	152,000
C(13, 0)	182,000

41. Refer to Exercise 12, Section 3.2. The problem is
    Minimize $C = 2x + 5y$ subject to
    $$30x + 25y \geq 400$$
    $$x + 0.5y \geq 10$$
    $$2x + 5y \geq 40$$
    $$x \geq 0, \ y \geq 0$$

The graph of the feasible set S and the associated table of values of C follow.

Vertex	$C = 2x + 5y$
A(0,20)	100
B(5,10)	60
C(10,4)	40
D(20,0)	40

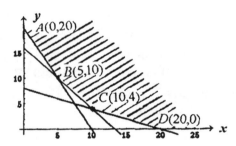

C attains a minimum value of 40 when $x =$
10 and $y = 4$ and $x = 20$, and $y = 0$. This means that any point lying on the line
joining the points (10,4) and (20,0) will satisfy these constraints. For example, we
could use 10 ounces of food A and 4 ounces of food B, or we could use 20 ounces
of food A and zero ounces of food B.

43. Refer to the solution of Exercise 13, Section 3.2. The problem is
    Maximize $P = 0.6x + 0.8y$ subject to

$$x + y \le 2.5$$
$$x + y \ge 2$$
$$1 \le x \le 1.5$$
$$y \ge 0.75$$

The feasible set S for the problem is shown in the figure at the right, and the value of P at each of the vertices of S is given in the table that follows.

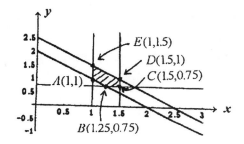

Vertex	$P = 0.6x + 0.8y$
A(1,1)	1.4
B(1.25,0.75)	1.35
C(1.5,0.75)	1.5
D(1.5,1)	1.7
E(1,1.5)	1.8

We conclude that the maximum value of P occurs when $x = 1$ and $y = 1.5$. Therefore, AntiFam should give $1 million in aid to country A and $1.5 million in aid to country B.

45. Refer to the solution of Exercise 15, Section 3.2.

Minimize $C = 32,000 - x - 3y$ subject to
$$x + y \le 6000$$
$$x + y \ge 2000$$
$$x \le 3000$$
$$y \le 4000$$
$$x \ge 0, \ y \ge 0$$

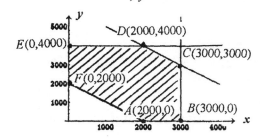

Vertex	$C = 32,000 - x - 3y$
A(2000,0)	30,000
B(3000,0)	29,000
C(3000,3000)	20,000
D(2000,4000)	18,000
E(0,4000)	20,000
F(0,2000)	26,000

Since x denotes the number of picture tubes shipped from location I to city A and y denotes the number of picture tubes shipped from location I to city B, we see that the company should ship 2000 tubes from location I to city A and 4000 tubes from location I to city B. Since the number of picture tubes required by the two factories in city A and city B are 3000 and 4000, respectively, the number of picture tubes shipped from location II to city A and city B, are

$(3000 - x) = 3000 - 2000 = 1000$ and $(4000 - y) = 4000 - 4000 = 0$

respectively. The minimum shipping cost will then be $18,000.

47. Let x denote Patricia's investment in growth stocks and y denote the value of her investment in speculative stocks, where both x and y are measured in thousands of dollars. Then the return on her investments is given by $P = 0.15x + 0.25y$. Since her investment may not exceed $30,000, we have the constraint $x + y \leq 30$. The condition that her investment in growth stocks be at least 3 times as much as her investment in speculative stocks translates into the inequality $x \geq 3y$. Thus, we have the following linear programming problem:

$$\text{Maximize } P = 0.15x + 0.25y \text{ subject to}$$
$$x + \ y \leq 30$$
$$x - 3y \geq \ 0$$
$$x \geq 0, \ y \geq 0$$

The graph of the feasible set S is shown in the figure and the value of P at each of the vertices of S is shown in the accompanying table. The maximum

Vertex	$C = 0.15x + 0.25y$
A(0,0)	0
B(30,0)	4.5
C($\frac{45}{2}$, $\frac{15}{2}$)	5.25

value of P occurs when $x = 22.5$ and $y = 7.5$. Thus, by investing $22,500 in growth stocks and $7,500 in speculative stocks. Patricia will realize a return of $5250 on her investments.

49. True. The optimal solution of a linear programming problem is the largest of all the feasible solutions.

51. False. The feasible set could be empty, and therefore, bounded.

53. Since the point $Q(x_1, y_1)$ lies in the interior of the feasible set S, it is possible to find another point $P(x_1, y_2)$ lying to the right and above the point Q and contained in S. (See figure.) Clearly, $x_2 > x_1$ and $y_2 > y_1$. Therefore, $ax_2 + by_2 > ax_1 + by_1$, since $a > 0$ and $b > 0$ and this shows that the objective function $P = ax + by$ takes on a larger value at P than it does at Q. Therefore, the optimal solution cannot occur at Q.

55. a.

b. There is no point that satisfies all the given inequalities. Therefore, there is no solution.

Exercises 3.4, page 212

1. Refer to the discussion on pages 203-204.
   a. Suppose the contribution to the profit of each type-B souvenir is $\$c$ so that
   $$P = x + cy, \text{ or } y = -\frac{x}{c} + \frac{p}{c}$$
   and the slope of the isoprofit line is $-1/c$. In order for the current optimal solution to be unaffected, this slope must be less than or equal to the slope of the line associated with constraint 2. Thus,
   $$-\frac{1}{c} \le -\frac{1}{3}, \quad \frac{1}{c} \ge \frac{1}{3} \text{ , or } c \le 3.$$
   Similarly, we see that the slope of the isoprofit line must be greater than or equal to the slope of the line associated with constraint 1. Thus

$$-\frac{1}{c} \ge -2, \quad \frac{1}{c} \le 2 \quad \text{or} \quad c \ge \frac{1}{2}.$$

Thus, the profit must lie between $0.50 and $3 as was to be shown.

b. If the profit of a type-A souvenir is $1.50, then the results on page 204 tell us that the current optimal solution holds. That is, we produce 48 type-A ($x = 48$) and 84 type B ($y = 84$) souvenirs giving a profit of
$$P = 1.5x + 1.2y = 1.5(48) + 1.2(84) = 172.8 \quad \text{or} \quad \$172.80.$$

c. Here the results of part (a) show that the current optimal solution holds. So, with $x = 48$ and $y = 84$, we find
$$P = x + 2y = 48 + 2(84) = 216 \quad \text{or} \quad \$216.$$

3. Refer to the discussion on pages 209-210.

a. Suppose resource 2 is changed from 1200 to 1200 + k. Then the constraint is changed to
$$6x + 3y = 1200 + k$$

The current optimal solution is shifted to the new optimal solution at the point D' (see the accompanying figure). To find the coordinates of D', we solve the system
$$3x + 4y = 1000$$
$$6x + 3y = 1200 + k$$

We obtain $x = \frac{4}{15}(450 + k)$ and $y = \frac{1}{5}(800 - k)$. The nonnegative requirements on x and y require that
$$450 + k \ge 0 \quad \text{or} \quad k \ge -450 \quad \text{and} \quad 800 - k \ge 0 \quad \text{or} \quad k \le 800.$$
Next, the constraint $x \le 180$ (constraint 3) requires that
$$\frac{4}{15}(450 + k) \le 180, \quad 450 + k \le 675, \quad \text{or} \quad k \le 225.$$
These three inequalities imply that $-450 \le k \le 225$. Therefore, resource 2 must lie between $1200 - 450$ and $1200 + 225$; that is, between 750 and 1425.

b. Suppose resource 3 is changed to $180 + \ell$. Refer to the accompanying figure.

To find the coordinates of D', we solve
$$6x + 3y = 1200$$
$$x \qquad = 180 + \ell$$

obtaining $x = 180 + \ell$ and $y = 40 - 2\ell$. Now for the current optimal solution to hold, $x = 180 + \ell \geq 120$ or $\ell \geq -60$. Therefore, resource 3 must be greater than or equal to $180 - 60 = 120$; that is, it cannot be decreased by more than 60.

5. a. The feasible set is shown in the accompanying figure. From the table

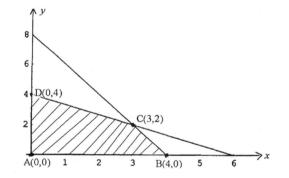

V	$P = 3x + 4y$
A(0,0)	0
B(4,0)	12
C(3,2)	$\boxed{17}$
D(0,4)	16

we see that the optimal solution is $x = 3, y = 2$, and $P = 17$.

b. Suppose $P = cx + 4y$. then $y = -\dfrac{c}{4}x + \dfrac{P}{4}$ and the slope of the isoprofit line is $-\dfrac{c}{4}$.
In order for the current optimal solution to hold, we must have

$$-\frac{c}{4} \leq -\frac{2}{3} \quad \text{or} \quad c \geq \frac{8}{3}. \quad \text{(Slope of constraint 1 is } -\frac{2}{3})$$

and $\quad -\dfrac{c}{4} \geq -2 \quad$ or $\quad c \leq 8. \quad$ (Slope of constraint 2 is -2)

Therefore $\frac{8}{3} \leq c \leq 8$

c. Suppose resource 1 is changed from 12 to 12 + h. Then the optimal solution is moved to a new optimal solution whose coordinates are found by solving the system

$$2x + 3y = 12 + h$$

$$2x + \ y = 8$$

The solutions are $x = 3 - \dfrac{h}{4}$ and $y = 2 + \dfrac{h}{2}$. Next, the nonnegativity of x and y imply that $h \ge -4$ and $h \le 12$. So, $-4 \le h \le 12$ and the resource can assume values between $12 - 4$ and $12 + 12$ or between 8 and 24.

d. If $h = 1$, then $x = 3 - \dfrac{1}{4}$ and $y = 2 + \dfrac{1}{2}$. The shadow price is

$$3[3 - \tfrac{1}{4}] + 4[2 + \tfrac{1}{2}] - [3(3) + 4(2)] = -\tfrac{3}{4} + 4(\tfrac{1}{2}) = \tfrac{5}{4}.$$

e. Since constraints 1 and 2 hold with equality at the optimal solution (3,2), they are both binding.

7.   a. The feasible set is shown in the accompanying figure.

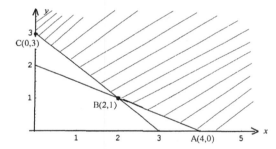

V	$P = 2x + 5y$
A(4,0)	8
B(2,1)	9
C(0,3)	15

From the table, we see that the optimal solution is $x = 4$, $y = 0$, and $C = 8$.

b. Suppose $C = cx + 5y$. then $y = -\dfrac{c}{5} + \dfrac{C}{5}$. On order for the current optimal solution to hold, we must have

$$-\frac{c}{5} \ge -\frac{1}{2}, \frac{c}{5} \le \frac{1}{2}, \ \text{or} \ c \le \frac{5}{2} \ \text{and} \ -\frac{c}{5} \ge -1, \frac{c}{5} \le 1, \ \text{or} \ c \le 5. \quad \text{Therefore,} \ 0 \le c \le \tfrac{5}{2}.$$

REMARK: For a minimization problem the optimal solution is the intersection of the corner point in the feasible set with the isoprofit line nearest the origin.

c. Suppose resource 1 is changed from 4 to 4 + h. Then solving the system

$$x + 2y = 4 + h$$

$$x + \ y = 3$$

we obtain $x = 2 - h$ and $y = 1 + h$. We must have $y = 1 + h \geq 0$ or $h \geq -1$ (see the figure). Therefore, resource 1 can assume values greater than or equal to $4 - 1$ or 3.

d. Put $h = 1$. Then we see that the shadow price for resource 1 is

$$[2(5) + 5(0)] - [2(4) + 5(0)] = 2 \qquad (\textit{Since } C = 2x + 5y)$$

e. Constraint 1 is binding since equality holds at the optimal solution, but constraint 2 is nonbinding.

9.   a. The feasible set is shown in the accompanying figure.

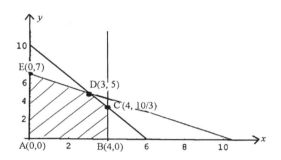

V	$P = 4x + 3y$
A(0,0)	0
B(4,0)	16
C(4, $\frac{10}{3}$)	26
D(3,5)	$\boxed{27}$
E(0,7)	21

From the table, we see that the optimal solution is $x = 3$, $y = 5$, and $P = 27$.

b. Suppose $P = cx + 3y$ so that $y = -\dfrac{c}{3}x + \dfrac{P}{3}$. For the current optimal solution to be unaffected, we must have

$$-\frac{c}{3} \geq -\frac{5}{3}, \quad \frac{c}{3} \leq \frac{5}{3}, \text{ or } c \leq 5$$

and

$$-\frac{c}{3} \leq -\frac{2}{3}, \quad \frac{c}{3} \geq \frac{2}{3}, \text{ or } c \geq 2. \text{ Therefore, } 2 \leq c \leq 5.$$

c. Suppose the right-hand side of the constraint is replaced by $30 + h$. Then we have the system

$$5x + 3y = 30 + h$$
$$2x + 3y = 21$$

The solutions are $x = 3 + \dfrac{h}{3}$ and $y = 5 - \dfrac{2h}{9}$. Since $x \geq 0$, we have $\dfrac{h}{3} + 3 \geq 0$ or $h \geq -9$. Also, $y \geq 0$ implies $5 - \dfrac{2h}{9} \geq 0$, $\dfrac{2h}{9} \leq 5$ or $h \leq \dfrac{45}{2}$. Finally, $x \leq 4$ implies $\dfrac{h}{3} + 3 \leq 4$ or $h \leq 3$. So $-9 \leq h \leq 3$ and we see that resource 1 can assume values between $30 - 9$ and $30 + 3$; that is, between 21 and 33.

d. Putting $h = 1$ in part (c) gives $x = 3 + \frac{1}{3} = \frac{10}{3}$ and $y - 5 - \frac{2}{9} = \frac{43}{9}$. Therefore, the shadow price associated with constraint 1 is

$$[4(\tfrac{10}{3}) + 3(\tfrac{43}{9})] - [4(3) + 3(5)] = \tfrac{2}{3}. \qquad \text{(Since } P = 4x + 3y)$$

e. The first two constraints are binding since equality holds at the optimal solution . Constraint 3 is nonbinding.

11. a. Let x and y denote the number of units of product A and the number of units of product B to be manufactured, respectively, Then the problem at hand is

$$\text{Maximize} \quad P = 3x + 4y$$
$$\text{subject to} \quad 6x + 9y \leq 300$$
$$5x + 4y \leq 180$$
$$x \geq 0, y \geq 0$$

The feasible set is shown in the accompanying figure. From the following table

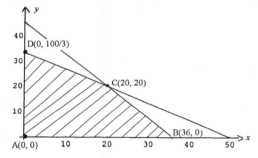

Vertex	$P = 3x + 4y$
A(0,0)	0
B(36,0)	108
C(20,20)	$\boxed{140}$
D(0,$133\frac{1}{3}$)	$133\frac{1}{3}$

we see that the optimal solution is $x = y = 20$, and $P = 140$. So, the company should produce 20 units of each product giving a maximum profit of $140.

b. Suppose $P = cx + 4y$ so that $y = -\frac{c}{4}x + \frac{P}{4}$. In order for the current optimal solution to remain optimal, we must have

$$-\frac{c}{4} \leq -\frac{2}{3}, \quad \frac{c}{4} \geq \frac{2}{3}, \quad \text{or} \quad c \geq \frac{8}{3} \quad \text{and} \quad -\frac{c}{4} \geq -\frac{5}{4}, \quad \frac{c}{4} \leq \frac{5}{4}, \quad \text{or} \quad c \leq 5.$$

Therefore, $\frac{8}{3} \leq c \leq 5$.

c. Suppose the right-hand side of the constraint is replaced by $300 + h$. Then we have the system

$$6x + 9y = 300 + h$$
$$5x + 4y = 180$$

The solutions are $x = 20 - \dfrac{4}{21}h$ and $y = 20 + \dfrac{5}{21}h$. Now, $x \geq 0$ implies

$20 - \frac{4}{21}h \geq 0$, or $h \leq 105$. Next, $y \geq 0$ implies $50 + \frac{5}{21}h \geq 0$, or $h \geq -84$.

Therefore, $-84 \leq h \leq 105$. So the resource can assume values between $300 - 84$ and $300 + 105$; that is, between 216 and 405.

d. Put $h = 1$ in part (c) and we find $x = 20 - \frac{4}{21}$, and $y = 20 + \frac{5}{21}h$. Using this result and the result from part (a), we see that the required shadow price is

$$3(20 - \tfrac{4}{21}) + 4(20 + \tfrac{5}{21}) - [3(20) + 4(20)] = \tfrac{8}{21}.$$

13.   a. Let x denote the number of days the Saddle Mine is operated, and let y denote the number of days the Horseshoe Mine is operated. Then the problem is

$$\text{Minimize } C = 14{,}000x + 16{,}000y$$
$$\text{subject to} \quad 50x + 75y \geq 650$$
$$3000x + 1000y \geq 18{,}000$$
$$x \geq 0, y \geq 0$$

The feasible set is shown in the accompanying figure. From the following table

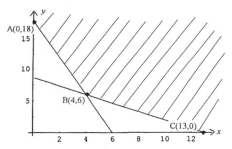

Vertex	$C = 14{,}000x + 16{,}000y$
A(0,18)	288,000
B(4,6)	152,000
C(13,0)	182,000

we see that the optimal solution is $x = 4$, $y = 6$, and $C = 152{,}000$. Therefore, the company should operate the Saddle Mine for 4 days and the Horseshoe Mine for 6 days at a cost of \$152,000.

b. Suppose $C = cx + 16{,}000y$. Then $y = -\dfrac{c}{16000}x + \dfrac{C}{16{,}000}$. In order for the current optimal solution to hold, we must have

$$-\frac{c}{16{,}000} \leq -\frac{50}{75}, \; \frac{c}{16{,}000} \geq \frac{50}{75}, \; \text{ or } \; c \geq 10{,}666\tfrac{2}{3}$$

and $-\dfrac{c}{16{,}000} \geq -3, \; \dfrac{c}{16{,}000} \leq 3$, or $c \leq 48{,}000$. Therefore, $10{,}666\frac{2}{3} \leq c \leq 48{,}000$.

c. Suppose the right-hand side of the first constraint (that pertains to the requirements for gold) is changed to $650 + h$. Then we have the system

$$50x + 75y = 650 + h$$
$$2000x + 1000y = 18,000$$

The solutions are $x = 4 - \frac{h}{175}$ and $y = 6 + \frac{3h}{175}$. Now, $x \geq 0$ implies

$4 - \frac{h}{175} \geq 0$, or $h \leq 700$. Next, $y \geq 0$ implies $6 + \frac{3h}{175} \geq 0$, or $h \geq -350$.

Therefore, $-350 \leq h \leq 700$ and so the resource can assume values between $650 - 350$ and $650 + 700$; that is, between 300 and 1350.

d. Put $h = 1$ in part (a), and we have $x = 4 - \frac{1}{175}$ and $y = 6 + \frac{3}{175}$. Using the fact that the optimal solution found in part (a) is $x = 4$ and $y = 6$, we see that the shadow price is

$$14,000(4 - \frac{1}{175}) + 16,000(6 + \frac{3}{175}) - [14,000(4) + 16,000(6)]$$
$$= 194.29, \text{ or } \$194.29.$$

15. a. Let x denote the number of Model A clock radios and y the number of Model B clock radios to be produced. We are led to the following problem.

$$\text{Maximize } P = 12x + 10y$$
$$\text{subject to } 15x + 10y \leq 1500$$
$$10x + 12y \leq 1320$$
$$x \leq 80$$
$$x \geq 0, y \geq 0$$

The feasible set is shown in the accompanying figure.

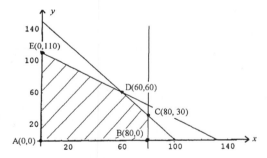

V	$P = 12x + 10y$
A(0,0)	0
B(80,0)	960
C(80,30)	1260
D(60,60)	1320
E(0,110)	1100

From the table we see that the optimal solution is $x = y = 60$, and $P = 1320$. So Soundex should produce 60 of each model and so that they realize a profit of $1320.

b. Suppose $P = cx + 10y$ so that $y = -\frac{c}{10}x + \frac{P}{10}$. In order for the current optimal solution to hold, we must have

$$-\frac{c}{10} \geq -\frac{3}{2}, \ \frac{c}{10} \leq \frac{3}{2}, \text{ or } c \leq 15 \text{ and } -\frac{c}{10} \leq -\frac{5}{6}, \ \frac{c}{10} \geq \frac{5}{6}, \text{ or } c \geq \frac{25}{3}.$$

Therefore, $8\frac{1}{3} \le c \le 15$.

c. Suppose the right-hand side of the first constraint (that pertains to the constraint on the use of Machine I) is changed to $1500 + h$. Then, we have the system

$$15x + 10y = 1500 + h$$

$$10x + 12y = 1320.$$

The solutions are $x = 60 + \frac{3}{20}h$ and $y = 60 - \frac{h}{8}$. Now, $x \ge 0$ implies $60 + \frac{3}{20}h \ge 0$, or $h \ge -400$. Next, $y \ge 0$ implies $60 - \frac{h}{8} \ge 0$ or $h \le 480$. Therefore, $-400 \le h \le \frac{400}{3}$ and we see that the values assumed by this resource must lie between $1500 - 400$ and $1500 + \frac{400}{3}$; that is, between 1100 and $1633\frac{1}{3}$.

d. Using the result of part (c) with $h = 1$, we find $x = 60 + \frac{3}{20}$ and $y = 60 - \frac{1}{8}$. Next, using the result of part (c), we see that the shadow price is

$$12(60 + \tfrac{3}{20}) + 10(60 - \tfrac{1}{8}) - [12(60) + 10(60)]$$

$$= 12(\tfrac{3}{20}) + 10(-\tfrac{1}{8}) = 0.55 \quad \text{or} \quad \$0.55.$$

e. Constraints 1 and 2 are binding since equality holds at the optimal solution. Constraint 3 is nonbinding.

17. a. Let x denote the number of model A grates and y the number of Model B grates to be produced. We are led to the following problem:

$$\text{Maximize} \quad P = 2x + 1.5y$$

$$\text{subject to} \quad 3x + 4y \le 1000 \quad \text{(Constraint 1)}$$

$$6x + 3y \le 1200 \quad \text{(Constraint 2)}$$

$$y \le 200 \quad \text{(Constraint 3)}$$

$$x \ge 0, y \ge 0$$

The feasible set is shown in the accompanying figure. From the table,

V	$P = 2x + 1.5y$
A(0,0)	0
B(200,0)	400
C(120,160)	$\boxed{480}$
D($\frac{200}{3}$,200)	433.33
E(0,200)	300

we see that the optimal solution is $x = 120$, $y = 160$, and $P = 480$. Therefore, Kane Manufacturing should produce 120 Model A grates and 160 Model B grates for a daily profit of $480.

b. Suppose $P = cx + \frac{3}{2}y$ so that $y = -\frac{2}{3}cx + \frac{2P}{3}$. In order for the current optimal solution to hold, we must have

$$-\frac{2c}{3} \leq -\frac{3}{4}, \quad \frac{2c}{3} \geq \frac{3}{4} \quad \text{or} \quad c \geq 1.125,$$

$$-\frac{2c}{3} \geq -2, \quad \frac{2c}{3} \leq 2 \quad \text{or} \quad c \leq 3,$$

and $\qquad -\frac{3}{2}c \leq 0 \quad$ or $\quad c \geq 0$. Therefore, $1.125 \leq c \leq 3$.

c. The constraint of interest here is constraint 1. Suppose we replace the right-hand side of this constraint by $1000 + h$. Then we have the following system.

$$3x + 4y = 1000 + h$$

$$6x + 3y = 1200$$

The solutions are $x = 120 - \frac{1}{5}h$ and $y = 160 + \frac{2}{5}h$. since $x \geq 0$, we have $120 - \frac{1}{5}h \geq 0$ or $h \leq 600$. Next, $y \geq 0$ implies $160 + \frac{2}{5}h \geq 0$ or $h \geq -400$. Finally, $y \leq 200$ implies $160 + \frac{2}{5}h \leq 200$, or $h \leq 100$. Therefore, $-400 \leq h \leq 100$. So, the values assumed by this resource must lie between $1000 - 400$ and $1000 + 100$; that is, between 600 and 1100.

d. Let $h = 1$. Then the results of part (c) give $x = 120 - \frac{1}{5}$ and $y = 160 + \frac{2}{5}$. Recall that the current optimal solution is $x = 120$ and $y = 160$. We see that the shadow price associated with the resource for cast iron (constraint 1) is

$$2(120 - \frac{1}{5}) + \frac{3}{2}(160 + \frac{2}{5}) - [2(120) + \frac{3}{2}(160)]$$

$$= 2(-\frac{1}{5}) + \frac{3}{2}(\frac{2}{5}) = \frac{1}{5} \quad \text{or} \quad \$0.20.$$

e. Constraints 1 and 2 are binding since equality holds at the optimal solution. Constraint 3 is nonbinding since constraint 3 reduces to $160 \leq 200$ when $x = 120$ and $y = 160$ and so equality does not hold.

CHAPTER 3 REVIEW EXERCISES, page 216

1. Evaluating Z at each of the corners of the feasible set S, we obtain the following table.

Vertex	$Z = 2x + 3y$
(0,0)	0
(5,0)	10
(3,4)	18
(0,6)	18

We conclude that Z attains a minimum value of 0 when $x = 0$ and $y = 0$, and a maximum value of 18 when x and y lie on the line segment joining $(3,4)$ and $(0,6)$.

3. The graph of the feasible set S is shown at the right.

Vertex	$Z = 3x + 5y$
$A(0,0)$	0
$B(5,0)$	15
$C(3,2)$	19
$D(0,4)$	20

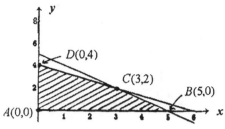

We conclude that the maximum value of P is 20 when $x = 0$ and $y = 4$.

5. The values of the objective function $C = 2x + 5y$ at the corners of the feasible set are given in the following table. The graph of the feasible set S follows.

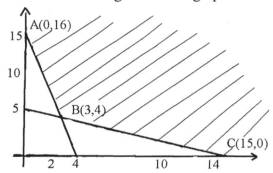

Vertex	$C = 2x + 5y$
$A(0,16)$	80
$B(3,4))$	$\boxed{26}$
$C(15,0)$	30

We conclude that the minimum value of C is 26 when $x = 3$ and $y = 4$.

7. The values of the objective function $P = 3x + 2y$ at the vertices of the feasible set are given in the following table. The graph of the feasible set is shown at the right.

Vertex	$P = 3x + 2y$
$A(0, \frac{28}{5})$	$\frac{56}{5}$
$B(7,0)$	21
$C(8,0)$	24
$D(3,10)$	$\boxed{29}$
$E(0,12)$	24

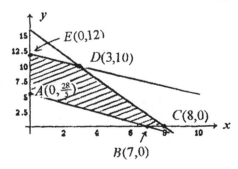

We conclude that P attains a maximum value of 29 when $x = 3$ and $y = 10$.

9. The graph of the feasible set S is shown at the right. The values of the objective function $C = 2x + 7y$ at each of the corner points of the feasible set S are shown in the table that follows.

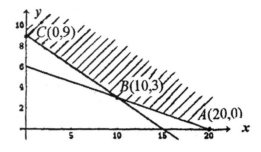

Vertex	$C = 2x + 7y$
A(20,0)	40
B(10,3)	41
C(0,9)	63

We conclude that C attains a minimum value of 40 when $x = 20$ and $y = 0$.

11. The graph of the feasible set S is shown in the following figure. We conclude that Q attains a maximum value of 22 when $x = 22$ and $y = 0$, and a minimum value of $5\frac{1}{2}$ when $x = 3$ and $y = \frac{5}{2}$.

Vertex	$Q = x + y$
A(2,5)	7
B(3,$\frac{5}{2}$)	$5\frac{1}{2}$
C(8,0)	8
D(22,0)	22

13. Suppose the investor puts x and y thousand dollars into the stocks of company A and company B, respectively. Then the mathematical formulation leads to the linear programming problem:

Maximize $P = 0.14x + 0.20y$ subject to
$$x + \quad y \le 80$$
$$0.01x + 0.04y \le 2$$
$$x \ge 0, \ y \ge 0$$

The feasible set S for this problem is shown in figure at the right and the values at each corner point are given in the accompanying table.

Vertex	$P = 0.14x + 0.20y$
A(0,0)	0
B(80,0)	11.2
C(40,40)	$\boxed{13.6}$
D(0,50)	10

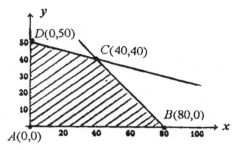

P attains a maximum value of 13.6 when $x = 40$ and $y = 40$. Thus, by investing $40,000 in the stocks of each company, the investor will achieve a maximum return of $13,600.

15. Let x denote the number of model A grates and y the number of model B grates to be produced. Then the constraint on the amount of cast iron available leads to
$$3x + 4y \le 1000$$
and the number of minutes of labor used each day leads to
$$6x + 3y \le 1200.$$
One additional constraint specifies that $y \ge 180$. The daily profit is $P = 2x + 1.5y$. Therefore, we have the following linear programming problem:

Maximize $P = 2x + 1.5y$ subject to
$$3x + 4y \le 1000$$
$$6x + 3y \le 1200$$
$$x \ge 0, \quad y \ge 180$$

The graph of the feasible set S is shown in the figure that follows.

Vertex	$P = 2x + 1.5y$
A(0,180)	270
B(0,250)	375
C($93\frac{1}{3}$,180)	$456\frac{2}{3}$

Thus, the optimal profit of $456 is realized when 93 units of model A grates and 180 units of model B grates are produced.

CHAPTER 4

EXERCISES 4.1, page 237

1. All entries in the last row of the simplex tableau are nonnegative and an optimal solution has been reached. We find
$$x = 30/7, \ y = 20/7, \ u = 0, \ v = 0, \ \text{and} \ P = 220/7.$$

3. The simplex tableau is not in final form because there is an entry in the last row that is negative. The entry in the first row, second column, is the next pivot element and has a value of $1/2$.

5. The simplex tableau is in final form. We find
$$x = 1/3, \ y = 0, \ z = 13/3, \ u = 0, \ v = 6, \ w = 0 \ \text{and} \ P = 17.$$

7. The simplex tableau is not in final form because there are two entries in the last row that are negative. The entry in the third row, second column, is the pivot element and has a value of 1.

9. The simplex tableau is in final form. The solutions are
$$x = 30, \ y = 0, \ z = 0, \ u = 10, \ v = 0, \ \text{and} \ P = 60,$$

and $x = 0, \ y = 30, \ z = 0, \ u = 10, \ v = 0, \ \text{and} \ P = 60.$
(There are infinitely many answers).

11. We obtain the following sequence of tableaus:

	x	y	u	v	P	Const
p.r. →	1	$\boxed{1}$	1	0	0	4
	2	1	0	1	0	5
	−3	−4	0	0	1	0

Ratio: 4, 5

$\xrightarrow{\substack{R_2 - R_1 \\ R_3 + 4R_1}}$

x	y	u	v	P	Const.
1	1	1	0	0	4
1	0	−1	1	0	1
1	0	4	0	1	16

p.c. (below second column of first tableau)

The last tableau is in final form and we conclude that $x = 0, \ y = 4, \ u = 0, \ v = 1,$ and $P = 16.$

4 Linear Programming: An Algebraic Approach **128**

13.

	x	y	u	v	P	Const		Ratio
p.r. →	1	[2]	1	0	0	12		12 / 2 = 6
	3	2	0	1	0	24		24 / 2 = 12
	−10	−12	0	0	1	0		

p.c. (under y)

$\xrightarrow{\tfrac{1}{2}R_1}$

	x	y	u	v	P	Const
	$\frac{1}{2}$	1	$\frac{1}{2}$	0	0	6
	3	2	0	1	0	24
	−10	−12	0	0	1	0

$\xrightarrow[R_3+12R_1]{R_2-2R_1}$

	x	y	u	v	P	Const		Ratio
	$\frac{1}{2}$	1	$\frac{1}{2}$	0	0	6		6 /(1 / 2) = 12
p.r. →	[2]	0	−1	1	0	12		12 / 2 = 6
	−4	0	6	0	1	72		

pc. (under x)

$\xrightarrow{\tfrac{1}{2}R_2}$

x	y	u	v	P	Const
$\frac{1}{2}$	1	$\frac{1}{2}$	0	0	6
1	0	$-\frac{1}{2}$	$\frac{1}{2}$	0	6
−4	0	6	0	1	72

$\xrightarrow[R_3+4R_2]{R_1-\tfrac{1}{2}R_2}$

x	y	u	v	P	Const
0	1	$\frac{3}{4}$	$-\frac{1}{4}$	0	3
1	0	$-\frac{1}{2}$	$\frac{1}{2}$	0	6
0	0	4	2	1	96

The last tableau is in final form. We find that $x = 6$, $y = 3$, $u = 0$, $v = 0$, and $P = 96$.

15. We obtain the following sequence of tableaus:

	x	y	u	v	w	P	Const.		Ratio
	3	1	1	0	0	0	24		24
	2	1	0	1	0	0	18		18
p.r. →	1	[3]	0	0	1	0	24		8
	−4	−6	0	0	0	1	0		

p.c. (under y)

	x	y	u	v	w	P	Const.
	3	1	1	0	0	0	24
$\xrightarrow{\frac{1}{3}R_3}$	2	1	0	1	0	0	18
	$\frac{1}{3}$	1	0	0	$\frac{1}{3}$	0	8
	-4	-6	0	0	0	1	0

$$\begin{array}{c} R_1-R_3 \\ R_2-R_3 \\ R_4+6R_3 \end{array} \longrightarrow$$

	x	y	u	v	w	P	Const.		Ratio
p.r.→	$\boxed{\frac{8}{3}}$	0	1	0	$-\frac{1}{3}$	0	16		6
	$\frac{5}{3}$	0	0	1	$-\frac{1}{3}$	0	10		6
	$\frac{1}{3}$	1	0	0	$\frac{1}{3}$	0	8		24
	-2	0	0	0	2	1	48		

$$\underset{\text{p.c.}}{\uparrow}$$

(Observe that we have a choice here.)

	x	y	u	v	w	P	Const.
	1	0	$\frac{3}{8}$	0	$-\frac{1}{8}$	0	6
$\xrightarrow{\frac{3}{8}R_1}$	$\frac{5}{3}$	0	0	1	$-\frac{1}{3}$	0	10
	$\frac{1}{3}$	1	0	0	$\frac{1}{3}$	0	8
	-2	0	0	0	2	1	48

$$\begin{array}{c} R_2-\frac{5}{3}R_1 \\ R_3-\frac{1}{3}R_1 \\ R_4+2R_1 \end{array} \longrightarrow$$

x	y	u	v	w	P	Const.
1	0	$\frac{3}{8}$	0	$-\frac{1}{8}$	0	6
0	0	$-\frac{5}{8}$	1	$-\frac{1}{8}$	0	0
0	1	$-\frac{1}{8}$	0	$\frac{3}{8}$	0	6
0	0	$\frac{3}{4}$	0	$\frac{7}{4}$	1	60

We deduce that $x = 6$, $y = 6$, $u = 0$, $v = 0$, $w = 0$, and $P = 60$.

17. We obtain the following sequence of tableaus:

	x	y	z	u	v	P	Const.
	1	1	1	1	0	0	8
p.r. →	3	2	$\boxed{4}$	0	1	0	24
	−3	−4	−5	0	0	1	0

Ratio
$8/1 = 8$
$24/4 = 6$ $\xrightarrow{\frac{1}{4}R_2}$

↑
p.c.

x	y	z	u	v	P	Const.
1	1	1	1	0	0	8
$\frac{3}{4}$	$\frac{1}{2}$	1	0	$\frac{1}{4}$	0	6
−3	−4	−5	0	0	1	0

$\xrightarrow[R_3+5R_2]{R_1-R_2}$

	x	y	z	u	v	P	Const.
p.r. →	$\frac{1}{4}$	$\boxed{\frac{1}{2}}$	0	1	$-\frac{1}{4}$	0	2
	$\frac{3}{4}$	$\frac{1}{2}$	1	0	$\frac{1}{4}$	0	6
	$\frac{3}{4}$	$-\frac{3}{2}$	0	0	$\frac{5}{4}$	1	30

Ratio
$2/(1/2) = 4$
$6/(1/2) = 12$ $\xrightarrow{2R_1}$

↑
p.c.

x	y	z	u	v	P	Const.
$\frac{1}{2}$	1	0	2	$-\frac{1}{2}$	0	4
$\frac{3}{4}$	$\frac{1}{2}$	1	0	$\frac{1}{4}$	0	6
$\frac{3}{4}$	$-\frac{3}{2}$	0	0	$\frac{5}{4}$	1	30

$\xrightarrow[R_3+\frac{3}{2}R_1]{R_2-\frac{1}{2}R_1}$

x	y	z	u	v	P	Const.
$\frac{1}{2}$	1	0	2	$-\frac{1}{2}$	0	4
$\frac{1}{2}$	0	1	−1	$\frac{1}{2}$	0	4
$\frac{3}{2}$	0	0	3	$\frac{1}{2}$	1	36

This last tableau is in final form. We find that $x = 0$, $y = 4$, $z = 4$, $u = 0$, $v = 0$, and $P = 36$.

19.

x	y	z	u	v	w	P	Const.
3	10	5	1	0	0	0	120
5	$\boxed{2}$	8	0	1	0	0	6
8	10	3	0	0	1	0	105
−3	−4	−1	0	0	0	1	0

p.r. → (row 2), p.c. ↑ (column y)

Ratio
$120/10 = 12$
$6/2 = 3$
$105/10 = 21/2$

$\xrightarrow{\frac{1}{2}R_2}$

x	y	z	u	v	w	P	Const.
3	10	5	1	0	0	0	120
$\frac{5}{2}$	1	4	0	$\frac{1}{2}$	0	0	3
8	10	3	0	0	1	0	105
−3	−4	−1	0	0	0	1	0

$\xrightarrow{\begin{array}{l} R_1-10R_2 \\ R_3-10R_2 \\ R_4+4R_2 \end{array}}$

x	y	z	u	v	w	P	Const.
−22	0	−35	1	−5	0	0	90
$\frac{5}{2}$	1	4	0	$\frac{1}{2}$	0	0	3
−17	0	−37	0	−5	1	0	75
7	0	15	0	2	0	1	12

The last tableau is in final form. We find that $x = 0$, $y = 3$, $z = 0$, $u = 90$, $v = 0$, $w = 75$, and $P = 12$.

21. We obtain the following sequence of tableaus:

x	y	z	u	v	w	P	Const.
1	1	1	1	0	0	0	20
2	$\boxed{4}$	3	0	1	0	0	42
2	0	3	0	0	1	0	30
−4	−6	−5	0	0	0	1	0

p.r. → (row 2), p.c. ↑ (column y)

Ratio
20
$10\frac{1}{2}$

$\xrightarrow{\frac{1}{4}R_2}$

x	y	z	u	v	w	P	Const.
1	1	1	1	0	0	0	20
$\frac{1}{2}$	1	$\frac{3}{4}$	0	$\frac{1}{4}$	0	0	$\frac{21}{2}$
2	0	3	0	0	1	0	30
-4	-6	-5	0	0	0	1	0

$$\xrightarrow{\substack{R_1-R_2\\R_4+6R_2}}$$

x	y	z	u	v	w	P	Const.	Ratio
$\frac{1}{2}$	0	$\frac{1}{4}$	1	$-\frac{1}{4}$	0	0	$\frac{19}{2}$	19
$\frac{1}{2}$	1	$\frac{3}{4}$	0	$\frac{1}{4}$	0	0	$\frac{21}{2}$	21
$\boxed{2}$	0	3	0	0	1	0	30	15
-1	0	$-\frac{1}{2}$	0	$\frac{3}{2}$	0	1	63	

p.r. → (row 3), p.c. ↑ (column x)

$$\xrightarrow{\frac{1}{2}R_3}$$

x	y	z	u	v	w	P	Const.
$\frac{1}{2}$	0	$\frac{1}{4}$	1	$-\frac{1}{4}$	0	0	$\frac{19}{2}$
$\frac{1}{2}$	1	$\frac{3}{4}$	0	$\frac{1}{4}$	0	0	$\frac{21}{2}$
1	0	$\frac{3}{2}$	0	0	$\frac{1}{2}$	0	15
-1	0	$-\frac{1}{2}$	0	$\frac{3}{2}$	0	1	63

$$\xrightarrow{\substack{R_1-\frac{1}{2}R_3\\R_2-\frac{1}{2}R_3\\R_4+R_3}}$$

x	y	z	u	v	w	P	Const.
0	0	$-\frac{1}{2}$	1	$-\frac{1}{4}$	$-\frac{1}{4}$	0	2
0	1	0	0	$\frac{1}{4}$	$-\frac{1}{4}$	0	3
1	0	$\frac{3}{2}$	0	0	$\frac{1}{2}$	0	15
0	0	1	0	$\frac{3}{2}$	$\frac{1}{2}$	1	78

So the solution is $x = 15$, $y = 3$, $z = 0$, $u = 2$, $v = 0$, $w = 0$, and $P = 78$.

23.   We obtain the following sequence of tableaus:

	x	y	z	u	v	w	P	Const.
p.r. →	$\boxed{2}$	1	1	1	0	0	0	10
	3	5	1	0	1	0	0	45
	2	5	1	0	0	1	0	40
	−12	−10	−5	0	0	0	1	0

↑
p.c.

Ratio
$10/2 = 5$
$45/3 = 15$ $\xrightarrow{\frac{1}{2}R_1}$
$40/2 = 20$

x	y	z	u	v	w	P	Const.
1	$\frac{1}{2}$	$\frac{1}{2}$	$\frac{1}{2}$	0	0	0	5
3	5	1	0	1	0	0	45
2	5	1	0	0	1	0	40
−12	−10	−5	0	0	0	1	0

$\begin{array}{c} R_2 - 3R_1 \\ R_3 - 2R_1 \\ R_4 + 12R_1 \end{array} \longrightarrow$

	x	y	z	u	v	w	P	Const.
	1	$\frac{1}{2}$	$\frac{1}{2}$	$\frac{1}{2}$	0	0	0	5
	0	$\frac{7}{2}$	$-\frac{1}{2}$	$-\frac{3}{2}$	1	0	0	30
p.r. →	0	$\boxed{4}$	0	−1	0	1	0	30
	0	−4	1	6	0	0	1	60

↑
.p.c

Ratio
$5/(1/2) = 10$
$30/(7/2) = 60/7$ $\xrightarrow{\frac{1}{4}R_3}$
$30/4 = 15/2$

x	y	z	u	v	w	P	Const.
1	$\frac{1}{2}$	$\frac{1}{2}$	$\frac{1}{2}$	0	0	0	5
0	$\frac{7}{2}$	$-\frac{1}{2}$	$-\frac{3}{2}$	1	0	0	30
0	1	0	$-\frac{1}{4}$	0	$\frac{1}{4}$	0	$\frac{15}{2}$
0	−4	1	6	0	0	1	60

$\begin{array}{c} R_1 - \frac{1}{2}R_3 \\ R_2 - \frac{7}{2}R_3 \\ R_4 + 4R_3 \end{array} \longrightarrow$

x	y	z	u	v	w	P	Const.
1	0	$\frac{1}{2}$	$\frac{5}{8}$	0	$-\frac{1}{8}$	0	$\frac{5}{4}$
0	0	$-\frac{1}{2}$	$-\frac{5}{8}$	1	$-\frac{7}{8}$	0	$\frac{15}{4}$
0	1	0	$-\frac{1}{4}$	0	$\frac{1}{4}$	0	$\frac{15}{2}$
0	0	1	5	0	1	1	90

This last tableau is in final form, and we conclude that $x = 5/4$, $y = 15/2$, $z = 0$, $u = 0$, $v = 15/4$, $w = 0$, and $P = 90$.

25. We obtain the following sequence of tableaus, where u, v and w are slack variables.

	x	y	z	u	v	w	P	Const.	Ratio	
p.r. →	[2]	1	2	1	0	0	0	7	$\frac{7}{2}$	
	2	3	1	0	1	0	0	8	4	$\frac{1}{2}R_1$ →
	1	2	3	0	0	1	0	7	7	
	−24	−16	−23	0	0	0	1	0		

p.c. ↑ (under x)

x	y	z	u	v	w	P	Const.	
[1]	$\frac{1}{2}$	1	$\frac{1}{2}$	0	0	0	$\frac{7}{2}$	$R_2 - 2R_1$
2	3	1	0	1	0	0	8	$R_3 - R_1$
1	2	3	0	0	1	0	7	$R_4 + 24R_1$ →
−24	−16	−23	0	0	0	1	0	

	x	y	z	u	v	w	P	Const.	Ratio	
	1	$\frac{1}{2}$	1	$\frac{1}{2}$	0	0	0	$\frac{7}{2}$	7	
p.r. →	0	[2]	−1	−1	1	0	0	1	$\frac{1}{2}$	$\frac{1}{2}R_2$ →
	0	$\frac{3}{2}$	2	$-\frac{1}{2}$	0	1	0	$\frac{7}{2}$	$\frac{7}{3}$	
	0	−4	1	12	0	0	1	84		

p.c. ↑ (under y)

x	y	z	u	v	w	P	Const.	
1	$\frac{1}{2}$	1	$\frac{1}{2}$	0	0	0	$\frac{7}{2}$	$R_1 - \frac{1}{2}R_2$
0	1	$-\frac{1}{2}$	$-\frac{1}{2}$	$\frac{1}{2}$	0	0	$\frac{1}{2}$	$R_3 - \frac{3}{2}R_2$
0	$\frac{3}{2}$	2	$-\frac{1}{2}$	0	1	0	$\frac{7}{2}$	$R_4 + 4R_2$ →
0	−4	1	12	0	0	1	84	

x	y	z	u	v	w	P	Const.		Ratio
1	0	$\frac{5}{4}$	$\frac{3}{4}$	$\frac{1}{4}$	0	0	$\frac{13}{4}$		$\frac{13}{5}$
0	1	$-\frac{1}{2}$	$-\frac{1}{2}$	$\frac{1}{2}$	0	0	$\frac{1}{2}$		$--$
p.r. → 0	0	$\boxed{\frac{11}{4}}$	$\frac{1}{4}$	$-\frac{3}{4}$	1	0	$\frac{11}{4}$		1
0	0	-1	10	2	0	1	86		

$\xrightarrow{\ \frac{4}{11}R_3\ }$

↑
p.c.

x	y	z	u	v	w	P	Const.
1	0	$\frac{5}{4}$	$\frac{3}{4}$	$\frac{1}{4}$	0	0	$\frac{13}{4}$
0	1	$-\frac{1}{2}$	$-\frac{1}{2}$	$\frac{1}{2}$	0	0	$\frac{1}{2}$
0	0	1	$\frac{1}{11}$	$-\frac{3}{11}$	$\frac{4}{11}$	0	1
0	0	-1	10	2	0	1	86

$\begin{array}{l} R_1-\frac{5}{4}R_3 \\ R_2+\frac{1}{2}R_3 \\ R_4+R_3 \end{array} \longrightarrow$

x	y	z	u	v	w	P	Const.
1	0	0	$\frac{7}{11}$	$\frac{13}{22}$	$-\frac{5}{11}$	0	2
0	1	0	$-\frac{5}{11}$	$\frac{4}{11}$	$\frac{2}{11}$	0	1
0	0	1	$\frac{1}{11}$	$-\frac{3}{11}$	$\frac{4}{11}$	0	1
0	0	0	$\frac{111}{11}$	$\frac{19}{11}$	$\frac{4}{11}$	1	87

This last tableau is in final form and we conclude that P attains a maximum value of 87 when $x = 2$, $y = 1$, and $z = 1$.

27. Pivoting about the x-column in the initial simplex tableau, we have

x	y	z	u	v	P	Const.		Ratio
3	3	-2	1	0	0	100		$100/3$
p.r. → $\boxed{5}$	5	3	0	1	0	150		$150/5$
-2	-2	4	0	0	1	0		

$\xrightarrow{\ \frac{1}{5}R_2\ }$

↑
p.c.

x	y	z	u	v	P	Const.	
3	3	-2	1	0	0	100	$\begin{array}{c} R_1-3R_2 \\ R_3+2R_2 \end{array} \longrightarrow$
1	1	$\frac{3}{5}$	0	$\frac{1}{5}$	0	30	
-2	-2	4	0	0	1	0	

x	y	z	u	v	P	Const.
0	0	$-\frac{19}{5}$	1	$-\frac{3}{5}$	0	10
1	1	$\frac{3}{5}$	0	$\frac{1}{5}$	0	30
0	0	$\frac{26}{5}$	0	$\frac{2}{5}$	1	60

and we see that one optimal solution occurs when $x = 30$, $y = 0$, $z = 0$, and $P = 60$. Similarly, pivoting about the y-column, we obtain another optimal solution: $x = 0$, $y = 30$, $z = 0$, and $P = 60$.

29. Let the number of model A and model B fax machines made each month be x and y, respectively. Then we have the following linear programming problem:

Maximize $P = 30x + 40y$ subject to

$$100x + 150y \le 600,000$$

$$x + y \le 2,500$$

$$x \ge 0, \ y \ge 0$$

Using the simplex method, we obtain the following sequence of tableaus:

x	y	u	v	P	Const	Ratio	
100	150	1	0	0	600,000	4000	$\begin{array}{c} R_1-150R_2 \\ R_3+40R_2 \end{array} \longrightarrow$
1	$\boxed{1}$	0	1	0	2,500	2500	
-30	-40	0	0	1	0		

x	y	u	v	P	Const
-50	0	1	-150	0	225000
1	1	0	1	0	2500
10	0	0	40	1	100000

We conclude that the maximum monthly profit is $100,000, and this occurs when 0 model A and 2500 model B fax machines are produced.

31. Suppose the farmer plants x acres of Crop A and y acres of crop B. Then the problem is

$$\text{Maximize } P = 150x + 200y \text{ subject to}$$
$$x + y \le 150$$
$$40x + 60y \le 7400$$
$$20x + 25y \le 3300$$
$$x \ge 0, \ y \ge 0$$

Using the simplex method, we obtain the following sequence of tableaus:

x	y	u	v	w	P	Const
1	1	1	0	0	0	150
40	60	0	1	0	0	7400
20	25	0	0	1	0	3300
−150	−200	0	0	0	1	0

$$\begin{array}{l} R_2 - 40R_1 \\ R_3 - 20R_1 \\ \underline{R_4 + 150R_1} \end{array} \longrightarrow$$

	x	y	u	v	w	P	Const	Ratio
	1	1	1	0	0	0	150	150
	0	20	−40	1	0	0	1400	700
$p.r. \to$	0	$\boxed{5}$	−20	0	1	0	300	60
	0	−50	150	0	0	1	22500	

$$\overset{\frac{1}{5}R_3}{\longrightarrow}$$

$$\underset{p.c.}{\uparrow}$$

x	y	u	v	w	P	Const
1	1	1	0	0	0	150
0	20	−40	1	0	0	1400
0	1	−4	0	$\frac{1}{5}$	0	60
0	−50	150	0	0	1	22500

$$\begin{array}{l} R_1 - R_3 \\ R_2 - 20R_3 \\ \underline{R_4 + 50R_3} \end{array} \longrightarrow$$

x	y	u	v	w	P	Const	Ratio
1	0	5	0	$-\frac{1}{5}$	0	90	18
0	0	$\boxed{40}$	1	-4	0	200	5
0	1	-4	0	$\frac{1}{5}$	0	60	$--$
0	0	-50	0	10	1	25500	

p.r. $\rightarrow$ (second row) $\xrightarrow{\frac{1}{40}R_2}$

p.c. (under u column)

x	y	u	v	w	P	Const
1	0	5	0	$-\frac{1}{5}$	0	90
0	0	1	$\frac{1}{40}$	$-\frac{1}{10}$	0	5
0	1	-4	0	$\frac{1}{5}$	0	60
0	0	-50	0	10	1	25500

$\xrightarrow[R_4+50R_2]{\substack{R_1-5R_2 \\ R_3+4R_2}}$

x	y	u	v	w	P	Const
1	0	0	$-\frac{1}{8}$	$\frac{3}{10}$	0	65
0	0	1	$\frac{1}{40}$	$-\frac{1}{10}$	0	5
0	1	0	$\frac{1}{10}$	$-\frac{1}{5}$	0	80
0	0	0	$\frac{5}{4}$	5	1	25750

The last tableau is in final form. We find $x = 65$, $y = 80$, and $P = 25,750$. So the maximum profit of \$25,750 is realized by planting 65 acres of Crop A and 80 acres of Crop B. Since $u = 5$, we see that there are 5 acres of land left unused.

33. We wish to maximize $P = 0.06x + 0.1y + 0.15z = \frac{3}{50}x + \frac{1}{10}y + \frac{3}{20}z$ subject to
$$x + y + z \le 250000$$
$$-x - y + 3z \le 0$$
$$-x + y - z \le 0$$
$$x \ge 0, y \ge 0, z \ge 0$$

Let u, v, and w be slack variables. We obtain the following tableas:

	x	y	z	u	v	w	P	Const.		Ratio	
p.r. →	1	1	1	1	0	0	0	250000		250000	
	-1	-1	3	0	1	0	0	0		0	$\frac{1}{3}R_2$ →
	-1	1	-1	0	0	1	0	0		--	
	$-\frac{3}{50}$	$-\frac{1}{10}$	$-\frac{3}{20}$	0	0	0	1	0			

$$\underset{p.c}{\uparrow}$$

x	y	z	u	v	w	P	Const.	
1	1	1	1	0	0	0	250000	
$-\frac{1}{3}$	$-\frac{1}{3}$	1	0	$\frac{1}{3}$	0	0	0	$\xrightarrow[\substack{R_1-R_2 \\ R_3+R_2 \\ R_4+\frac{3}{20}R_2}]{}$
-1	1	-1	0	0	1	0	0	
$-\frac{3}{50}$	$-\frac{1}{10}$	$-\frac{3}{20}$	0	0	0	1	0	

x	y	z	u	v	w	P	Const.		Ratio	
$\frac{4}{3}$	$\frac{4}{3}$	0	1	$-\frac{1}{3}$	0	0	250000		$\frac{250000}{4/3}=187500$	
$-\frac{1}{3}$	$-\frac{1}{3}$	1	0	$\frac{1}{3}$	0	0	0		--	$\frac{3}{2}R_3$ →
$-\frac{4}{3}$	$\frac{2}{3}$	0	0	$\frac{1}{3}$	1	0	0		0	
$-\frac{11}{100}$	$-\frac{3}{20}$	0	0	$\frac{1}{20}$	0	1	0			

$$\underset{p.c.}{\uparrow}$$

x	y	z	u	v	w	P	Const.	
$\frac{4}{3}$	$\frac{4}{3}$	0	1	$-\frac{1}{3}$	0	0	250000	
$-\frac{1}{3}$	$-\frac{1}{3}$	1	0	$\frac{1}{3}$	0	0	0	$\xrightarrow[\substack{R_1-\frac{4}{3}R_3 \\ R_2+\frac{1}{3}R_3 \\ R_4+\frac{3}{20}R_2}]{}$
-2	1	0	0	$\frac{1}{2}$	$\frac{3}{2}$	0	0	
$-\frac{11}{100}$	$-\frac{3}{20}$	0	0	$\frac{1}{20}$	0	1	0	

x	y	z	u	v	w	P	Const.	Ratio
$\boxed{4}$	0	0	1	-1	-2	0	250000	62500
-1	0	1	0	$\frac{1}{2}$	$\frac{1}{2}$	0	0	$--$
-2	1	0	0	$\frac{1}{2}$	$\frac{3}{2}$	0	0	0
$-\frac{41}{100}$	0	0	0	$\frac{3}{20}$	$\frac{9}{40}$	1	0	

$\xrightarrow{\frac{1}{4}R_1}$

p.c. ↑

x	y	z	u	v	w	P	Const.
$\boxed{1}$	0	0	$\frac{1}{4}$	$-\frac{1}{4}$	$-\frac{1}{2}$	0	62500
-1	0	1	0	$\frac{1}{2}$	$\frac{1}{2}$	0	0
-2	1	0	0	$\frac{1}{2}$	$\frac{3}{2}$	0	0
$-\frac{41}{100}$	0	0	0	$\frac{3}{20}$	$\frac{9}{40}$	1	0

$\xrightarrow[\substack{R_3+2R_2 \\ R_4+\frac{41}{100}R_1}]{R_2+R_1}$

x	y	z	u	v	w	P	Const.
1	0	0	$\frac{1}{4}$	$-\frac{1}{4}$	$-\frac{1}{2}$	0	62500
0	0	1	$\frac{1}{4}$	$\frac{1}{4}$	0	0	62500
0	1	0	$\frac{1}{2}$	0	$\frac{1}{2}$	0	125000
0	0	0	$\frac{41}{400}$	$\frac{19}{400}$	$\frac{1}{50}$	1	25,625

The last ableau is in final form, and we see that $x = 62,500$, $y = 125,000$, $z = 62,500$, and $P = 25,625$. So, Ashley should invest \$62,500 in the money market fund, \$125,000 in the international equity fund, and \$62,500 in the growth-and-income fund. Her maximum return will be \$25,625.

35. Suppose the Excelsior Company buys x, y, and z minutes of morning, afternoon, and evening commercials, respectively. Then we wish to maximize
$$P = 200,000x + 100,000y + 600,000z \text{ subject to}$$
$$3000x + 1000y + 12,000z \le 102,000$$
$$z \le 6$$
$$x + y + z \le 25$$
$$x \ge 0, y \ge 0, z \ge 0$$

Using the simplex method, we obtain the following sequence of tableaus.

x	y	z	u	v	w	P	Const.	Ratio
$\boxed{3000}$	1000	12,000	1	0	0	0	102,000	17/2
0	0	1	0	1	0	0	6	6
1	1	1	0	0	1	0	25	25
$-200,000$	$-100,000$	$-600,000$	0	0	0	1	0	

$\xrightarrow{\begin{array}{l} R_1-12,000R_2 \\ R_3-R_2 \\ R_4+600,000R_2 \end{array}}$

x	y	z	u	v	w	P	Const.	Ratio
3000	1000	0	1	$-12,000$	0	0	30,000	10
0	0	1	0	1	0	0	6	$--$
1	1	0	0	-1	1	0	19	19
$-200,000$	$-100,000$	0	0	600,000	0	1	3,600,000	

$\xrightarrow{\frac{1}{3000}R_1}$

x	y	z	u	v	w	P	Const.
1	$\frac{1}{3}$	0	$\frac{1}{3000}$	-4	0	0	10
0	0	1	0	1	0	0	6
1	1	0	0	-1	1	0	19
$-200,000$	$-100,000$	0	0	600,000	0	1	3,600,000

$\xrightarrow{\begin{array}{l} R_3-R_1 \\ R_4+200,000R_1 \end{array}}$

x	y	z	u	v	w	P	Const.
1	$\frac{1}{3}$	0	$\frac{1}{3000}$	-4	0	0	10
0	0	1	0	1	0	0	6
0	$\frac{2}{3}$	0	$-\frac{1}{3000}$	$\boxed{3}$	1	0	9
0	$-\frac{100,000}{3}$	0	$\frac{200}{3}$	$-200,000$	0	1	5,600,000

$\xrightarrow{\frac{1}{3}R_3}$

x	y	z	u	v	w	P	Const.
1	$\frac{1}{3}$	0	$\frac{1}{3000}$	-4	0	0	10
0	0	1	0	1	0	0	6
0	$\frac{2}{9}$	0	$-\frac{1}{9000}$	1	$\frac{1}{3}$	0	3
0	$-\frac{100,000}{3}$	0	$\frac{200}{3}$	$-200,000$	0	1	5,600,000

$\xrightarrow{\begin{array}{l} R_1+4R_3 \\ R_2-R_3 \\ R_4+200,000R_3 \end{array}}$

x	y	z	u	v	w	P	Const.
1	$\frac{11}{9}$	0	$-\frac{1}{9000}$	0	$\frac{4}{3}$	0	22
0	$-\frac{2}{9}$	1	$\frac{1}{9000}$	0	$-\frac{1}{3}$	0	3
0	$\frac{2}{9}$	0	$-\frac{1}{9000}$	1	$\frac{1}{3}$	0	3
0	$\frac{100,000}{9}$	0	$\frac{400}{9}$	0	$\frac{200,000}{3}$	1	6,200,000

We conclude that $x = 22$, $y = 0$, $z = 3$, $u = 0$, $v = 3$, and $P = 6,200,000$. Therefore, the company should buy 22 minutes of morning and 3 minutes of evening advertising time, thereby maximizing their exposure to 6,200,000 viewers.

37. We first tabulate the given information:

Department	Model A	Model B	Model C	Time available
Fabrication	$\frac{5}{4}$	$\frac{3}{2}$	$\frac{3}{2}$	310
Assembly	1	1	$\frac{3}{4}$	205
Finishing	1	1	$\frac{1}{2}$	190
Profit	26	28	24	

Let x, y, and z denote the number of units of model A, model B, and model C to be produced, respectively. Then the required linear programming problem is

Maximize $P = 26x + 28y + 24z$ subject to

$$\frac{5}{4}x + \frac{3}{2}y + \frac{3}{2}z \leq 310$$
$$x + y + \frac{3}{4}z \leq 205$$
$$x + y + \frac{1}{2}z \leq 190$$
$$x \geq 0, y \geq 0, z \geq 0$$

Using the simplex method, we obtain the following tableaus:

	x	y	z	u	v	w	P	Const.		Ratio	
	$\frac{5}{4}$	$\frac{3}{2}$	$\frac{3}{2}$	1	0	0	0	310		$206\frac{2}{3}$	$R_1 - \frac{3}{2}R_3$
	1	1	$\frac{3}{4}$	0	1	0	0	205		205	$R_2 - R_3$
p.r. →	1	$\boxed{1}$	$\frac{1}{2}$	0	0	1	0	190		190	$\xrightarrow{R_4 + 28R_3}$
	-26	-28	-24	0	0	0	1	0			
		$\uparrow$									
		p.c.									

	x	y	z	u	v	w	P	Const.		Ratio	
p.r. →	$-\frac{1}{4}$	0	$\boxed{\frac{3}{4}}$	1	0	$-\frac{3}{2}$	0	25		$33\frac{1}{3}$	
	0	0	$\frac{1}{4}$	0	1	-1	0	15		60	$\xrightarrow{\frac{4}{3}R_1}$
	1	1	$\frac{1}{2}$	0	0	1	0	190		380	
	2	0	-10	0	0	28	1	5320			
			$\uparrow$								
			p.c.								

	x	y	z	u	v	w	P	Const.	
	$-\frac{1}{3}$	0	1	$\frac{4}{3}$	0	-2	0	$\frac{100}{3}$	$R_2 - \frac{1}{4}R_1$
	0	0	$\frac{1}{4}$	0	1	-1	0	15	$R_3 - \frac{1}{2}R_1$
	1	1	$\frac{1}{2}$	0	0	1	0	190	$\xrightarrow{R_4 + 10R_1}$
	2	0	-10	0	0	28	1	5320	

	x	y	z	u	v	w	P	Const.		Ratio	
	$-\frac{1}{3}$	0	1	$\frac{4}{3}$	0	-2	0	$\frac{100}{3}$		$--$	
p.r. →	$\boxed{\frac{1}{12}}$	0	0	$-\frac{1}{3}$	1	$-\frac{1}{2}$	0	$\frac{20}{3}$		80	$\xrightarrow{12R_2}$
	$\frac{7}{6}$	1	0	$-\frac{2}{3}$	0	2	0	$\frac{520}{3}$		$148\frac{4}{7}$	
	$-\frac{4}{3}$	0	0	$\frac{40}{3}$	0	8	1	$\frac{16960}{3}$			
	$\uparrow$										
	p.c.										

x	y	z	u	v	w	P	Const.
$-\frac{1}{3}$	0	1	$\frac{4}{3}$	0	-2	0	$\frac{100}{3}$
1	0	0	-4	12	-6	0	80
$\frac{7}{6}$	1	0	$-\frac{2}{3}$	0	2	0	$\frac{520}{3}$
$-\frac{4}{3}$	0	0	$\frac{40}{3}$	0	8	1	$\frac{16960}{3}$

$R_1 + \frac{1}{3}R_2$
$R_3 - \frac{7}{6}R_2$
$R_4 + \frac{4}{3}R_2$
$\longrightarrow$

x	y	z	u	v	w	P	Const.
0	0	1	0	4	-4	0	60
1	0	0	-4	12	-6	0	80
0	1	0	4	-14	9	0	80
0	0	0	8	16	0	1	5760

The last tableau is in final form. We see that $x = 80$, $y = 80$, $z = 60$, $u = 0$, $v = 0$, $w = 0$, and $P = 5760$. So, by producing 80 units each of Models A and B, and 60 units of Model C, the company stands to make a profit of $5760. Since $u = v = w = 0$, there are no resources left over.

39. Let x, y, and z denote the number (in thousands) of bottles of formula I, formula II, and formula III, respectively, produced. The resulting linear programming problem is

$$\text{Maximize } P = 180x + 200y + 300z \text{ subject to}$$
$$\tfrac{5}{2}x + 3y + 4z \le 70$$
$$x \le 9$$
$$y \le 12$$
$$z \le 6$$
$$x \ge 0,\ y \ge 0,\ z \ge 0$$

Using the simplex method, we have

	x	y	z	s	t	u	v	P	Const.	Ratio
	$\frac{5}{2}$	3	4	1	0	0	0	0	70	$17\frac{1}{2}$
	1	0	0	0	1	0	0	0	9	$--$
	0	1	0	0	0	1	0	0	12	$--$
p.r. $\rightarrow$	0	0	$\boxed{1}$	0	0	0	1	0	6	6
	-180	-200	-300	0	0	0	0	1	0	

$R_1 - 4R_4$
$R_5 + 300R_4$
$\longrightarrow$

p.c.

	x	y	z	s	t	u	v	P	Const.		Ratio
	$\frac{5}{2}$	3	0	1	0	0	-4	0	46		$15\frac{1}{3}$
	1	0	0	0	1	0	0	0	9		--
p.r. $\rightarrow$	0	$\boxed{1}$	0	0	0	1	0	0	12		12
	0	0	1	0	0	0	1	0	6		--
	-180	-200	0	0	0	0	300	1	1800		

$$\underset{\text{p.c.}}{\uparrow}$$

$$\xrightarrow{\begin{array}{c} R_1-3R_3 \\ R_5+200R_3 \end{array}}$$

x	y	z	s	t	u	v	P	Const.		Ratio
$\frac{5}{2}$	0	0	1	0	-3	-4	0	10		4
1	0	0	0	1	0	0	0	9		9
0	1	0	0	0	1	0	0	12		--
0	0	1	0	0	0	1	0	6		--
-180	0	0	0	0	200	300	1	4200		

$$\xrightarrow{\frac{2}{5}R_1}$$

x	y	z	s	t	u	v	P	Const.
1	0	0	$\frac{2}{5}$	0	$-\frac{6}{5}$	$-\frac{8}{5}$	0	4
1	0	0	0	1	0	0	0	9
0	1	0	0	0	1	0	0	12
0	0	1	0	0	0	1	0	6
-180	0	0	0	0	200	300	1	4200

$$\xrightarrow{\begin{array}{c} R_2-R_1 \\ R_5+180R_1 \end{array}}$$

x	y	z	s	t	u	v	P	Const.		Ratio
1	0	0	$\frac{2}{5}$	0	$-\frac{6}{5}$	$-\frac{8}{5}$	0	4		--
1	0	0	$-\frac{2}{5}$	1	$\boxed{\frac{6}{5}}$	$\frac{8}{5}$	0	5		$\frac{25}{6}$
0	1	0	0	0	1	0	0	12		12
0	0	1	0	0	0	1	0	6		--
0	0	0	72	0	-16	12	1	4920		

$$\xrightarrow{\frac{5}{6}R_2}$$

x	y	z	s	t	u	v	P	Const.
1	0	0	$\frac{2}{5}$	0	$-\frac{6}{5}$	$-\frac{8}{5}$	0	4
0	0	0	$-\frac{1}{3}$	$\frac{5}{6}$	1	$\frac{4}{3}$	0	$\frac{25}{6}$
0	1	0	0	0	1	0	0	12
0	0	1	0	0	0	1	0	6
0	0	0	72	0	-16	12	1	4920

$R_1 + \frac{6}{5}R_2$
$R_3 - R_2$
$R_5 + 16R_2 \longrightarrow$

x	y	z	s	t	u	v	P	Const.
1	0	0	0	1	0	0	0	9
0	0	0	$-\frac{1}{3}$	$\frac{5}{6}$	1	$\frac{4}{3}$	0	$\frac{25}{6}$
0	1	0	$\frac{1}{3}$	$-\frac{5}{6}$	0	$-\frac{4}{3}$	0	$\frac{47}{6}$
0	0	1	0	0	0	1	0	6
0	0	0	$\frac{200}{3}$	$\frac{40}{3}$	0	$\frac{100}{3}$	1	$4986\frac{2}{3}$

Therefore, $x = 9$, $y = 47/6$, $z = 6$, $s = 0$, $t = 0$, $u = \frac{25}{6}$ and $P \approx 4986.67$; that is, the company should manufacture 9000 bottles of formula I, 7833 bottles of formula II, and 6000 bottles of formula III for a maximum profit of $4986.67.

41. Refer to Exercise 11, page 203. Let u, v, w, and s be slack variables. We obtain the following tableaus.

	x	y	z	u	v	w	s	P	Const.		Ratio
	1	1	1	1	0	0	0	0	2000000		2000000
$p.r. \rightarrow$	-2	-2	$\boxed{8}$	0	1	0	0	0	0		0
	-6	4	4	0	0	1	0	0	0		0
	-10	6	6	0	0	0	1	0	0		0
	$-\frac{1}{10}$	$-\frac{3}{20}$	$-\frac{1}{5}$	0	0	0	0	1	0		

$\frac{1}{8}R_2 \longrightarrow$

$\uparrow$
$p.c.$

x	y	z	u	v	w	s	P	Const.	
1	1	1	1	0	0	0	0	2000000	
$-\frac{1}{4}$	$-\frac{1}{4}$	$\boxed{1}$	0	$\frac{1}{8}$	0	0	0	0	$\xrightarrow{\begin{array}{c}R_1-R_2\\R_3-4R_2\\R_4-6R_2\\R_5+\frac{1}{5}R_2\end{array}}$
-6	4	4	0	0	1	0	0	0	
-10	6	6	0	0	0	1	0	0	
$-\frac{1}{10}$	$-\frac{3}{20}$	$-\frac{1}{5}$	0	0	0	0	1	0	

x	y	z	u	v	w	s	P	Const.	Ratio	
$\frac{5}{4}$	$\frac{5}{4}$	0	1	$-\frac{1}{8}$	0	0	0	2000000	1600000	
$-\frac{1}{4}$	$-\frac{1}{4}$	1	0	$\frac{1}{8}$	0	0	0	0	$--$	$\xrightarrow{\frac{1}{5}R_3}$
-5	$\boxed{5}$	0	0	$-\frac{1}{2}$	1	0	0	0	0	
$-\frac{17}{2}$	$\frac{15}{2}$	0	0	$-\frac{3}{4}$	0	1	0	0	0	
$-\frac{3}{20}$	$-\frac{1}{5}$	0	0	$\frac{1}{40}$	0	0	1	0		

x	y	z	u	v	w	s	P	Const.	
$\frac{5}{4}$	$\frac{5}{4}$	0	1	$-\frac{1}{8}$	0	0	0	2000000	
$-\frac{1}{4}$	$-\frac{1}{4}$	1	0	$\frac{1}{8}$	0	0	0	0	
-1	$\boxed{1}$	0	0	$-\frac{1}{10}$	$\frac{1}{5}$	0	0	0	$\xrightarrow{\begin{array}{c}R_1-\frac{5}{4}R_3\\R_2+\frac{1}{4}R_3\\R_4-\frac{15}{2}R_3\\R_5+\frac{1}{5}R_3\end{array}}$
$-\frac{17}{2}$	$\frac{15}{2}$	0	0	$-\frac{3}{4}$	0	1	0	0	
$-\frac{3}{20}$	$-\frac{1}{5}$	0	0	$\frac{1}{40}$	0	0	1	0	

x	y	z	u	v	w	s	P	Const.	Ratio	
$\boxed{\frac{5}{2}}$	0	0	1	$-\frac{1}{4}$	0	0	0	2000000	800000	
$-\frac{1}{2}$	0	1	0	$\frac{1}{10}$	$\frac{1}{20}$	0	0	0	$--$	$\xrightarrow{\frac{2}{5}R_1}$
-1	1	0	0	$-\frac{1}{10}$	$\frac{1}{5}$	0	0	0	0	
-1	0	0	0	0	$-\frac{3}{2}$	1	0	0	0	
$-\frac{7}{20}$	0	0	0	$\frac{1}{200}$	$\frac{1}{25}$	0	1	0		

x	y	z	u	v	w	s	P	Const.
$\boxed{1}$	0	0	$\frac{2}{5}$	0	$-\frac{1}{10}$	0	0	800000
$-\frac{1}{2}$	0	1	0	$\frac{1}{10}$	$\frac{1}{20}$	0	0	0
-1	1	0	0	$-\frac{1}{10}$	$\frac{1}{5}$	0	0	0
-1	0	0	0	0	$-\frac{3}{2}$	1	0	0
$-\frac{7}{20}$	0	0	0	$\frac{1}{200}$	$\frac{1}{25}$	0	1	0

$$\xrightarrow{\begin{array}{l} R_2+\frac{1}{2}R_1 \\ R_3+R_1 \\ R_4+R_1 \\ R_5+\frac{7}{20}R_1 \end{array}}$$

x	y	z	u	v	w	s	P	Const.
1	0	0	$\frac{2}{5}$	0	$-\frac{1}{10}$	0	0	800000
0	0	1	$\frac{1}{5}$	$\frac{1}{10}$	0	0	0	400000
0	1	0	$\frac{2}{5}$	$-\frac{1}{10}$	$\frac{1}{10}$	0	0	800000
0	0	0	$\frac{2}{5}$	0	$-\frac{8}{5}$	1	0	800000
0	0	0	$\frac{7}{50}$	$\frac{1}{200}$	$\frac{1}{200}$	0	1	280000

The last tableau is in final form, and we see that $x = 800,000$, $y = 800,000$, $Z = 400,000$, and $P = 280,000$. Thus, the financier should invest \$800,000 each in Projects A and B, and \$400,000 in project C. The maximum returns are \$280,000.

43. False. consider the linear programming problem

Maximize $P = 2x + 3y$ subject to
$$-x + y \leq 0$$
$$x \geq 0, \ y \geq 0$$

45. True. Consider the objective function $P = c_1 x + c_2 x_2 + \cdots c_n x_n$, which may be written in the form
$$-c_1 x_1 - c_2 x_2 - \cdots - c_n x_n + P = 0$$
Observe that the most negative of the numbers $-c_1, -c_2, \cdots, -c_n$ (which are the numbers comprising the last row of the simplex tableau) is just the largest coefficient of x_i in the expression for P. Thus, moving in the direction of the variable with this coefficient ensures that P increases most.

1. $x = 1.2, y = 0, z = 1.6, w = 0,$ and $P = 8.8$

3. $x = 1.6, y = 0, z = 0, w = 3.6,$ and $P = 12.4$

EXERCISES 4.2, page 257

1.   We solve the associated regular problem:

Maximize $P = -C = 2x - y$ subject to
$$x + 2y \leq 6$$
$$3x + 2y \leq 12$$
$$x \geq 0, \ y \geq 0$$

Using the simplex method where u and v are slack variables, we have

	x	y	u	v	P	Const.	Ratio	
	1	2	1	0	0	6	6	$\frac{1}{3}R_2$
p.r.→	3	2	0	1	0	12	4	→
	−2	1	0	0	1	0		

↑

p.c.

x	y	u	v	P	Const.
1	2	1	0	0	6
1	$\frac{2}{3}$	0	$\frac{1}{3}$	0	4
−2	1	0	0	1	0

$$\xrightarrow[R_3+2R_2]{R_1-R_2}$$

x	y	u	v	P	Const.
0	$\frac{4}{3}$	1	$-\frac{1}{3}$	0	2
1	$\frac{2}{3}$	0	$\frac{1}{3}$	0	4
0	$\frac{7}{3}$	0	$\frac{2}{3}$	1	8

Therefore, $x = 4, y = 0,$ and $C = -P = -8$.

3.   We maximize $P = -C = 3x + 2y$. Using the simplex method, we obtain

	x	y	u	v	P	Const.	Ratio	
	3	4	1	0	0	24	8	$\frac{1}{7}R_2$
p.r.→	7	−4	0	1	0	16	$\frac{16}{7}$	→
	−3	−2	0	0	1	0		

↑
p.c.

x	y	u	v	P	Const.
3	4	1	0	0	24
1	$-\frac{4}{7}$	0	$\frac{1}{7}$	0	$\frac{16}{7}$
-3	-2	0	0	1	0

$\xrightarrow{\substack{R_1-3R_2 \\ R_3+3R_2}}$

	x	y	u	v	P	Const.	Ratio
p.r. →	0	$\boxed{\frac{40}{7}}$	1	$-\frac{3}{7}$	0	$\frac{120}{7}$	3
	1	$-\frac{4}{7}$	0	$\frac{1}{7}$	0	$\frac{16}{7}$	$--$
	0	$-\frac{26}{7}$	0	$\frac{3}{7}$	1	$\frac{48}{7}$	

↑
p.c.

$\xrightarrow{\frac{7}{40}R_1}$

x	y	u	v	P	Const.
0	1	$\frac{7}{40}$	$-\frac{3}{40}$	0	3
1	$-\frac{4}{7}$	0	$\frac{1}{7}$	0	$\frac{16}{7}$
0	$-\frac{26}{7}$	0	$\frac{3}{7}$	1	$\frac{48}{7}$

$\xrightarrow{\substack{R_2+\frac{4}{7}R_1 \\ R_3+\frac{26}{7}R_1}}$

x	y	u	v	P	Const.
0	1	$\frac{7}{40}$	$-\frac{3}{40}$	0	3
1	0	$\frac{1}{10}$	$\frac{1}{10}$	0	4
0	0	$\frac{13}{20}$	$\frac{3}{20}$	1	18

The last tableau is in final form. We find $x = 4$, $y = 3$, and $C = -P = -18$.

5.   We maximize $P = -C = -2x + 3y + 4z$ subject to the given constraints. Using the simplex method we obtain

	x	y	z	u	v	w	P	Const.	Ratio
	-1	2	-1	1	0	0	0	8	$--$
p.r. →	1	-2	$\boxed{2}$	0	1	0	0	10	5
	2	4	-3	0	0	1	0	12	$--$
	2	-3	-4	0	0	0	1	0	

↑
p.c.

$\xrightarrow{\frac{1}{2}R_2}$

x	y	z	u	v	w	P	Const.
-1	2	-1	1	0	0	0	8
$\frac{1}{2}$	-1	1	0	$\frac{1}{2}$	0	0	5
2	4	-3	0	0	1	0	12
2	-3	-4	0	0	0	1	0

$\xrightarrow{\substack{R_1+R_2 \\ R_3+3R_2 \\ R_4+4R_2}}$

	x	y	z	u	v	w	P	Const.	Ratio	
$p.r. \rightarrow$	$-\frac{1}{2}$	$\boxed{1}$	0	1	$\frac{1}{2}$	0	0	13	13	$R_2 + R_1$
	$\frac{1}{2}$	-1	1	0	$\frac{1}{2}$	0	0	5	$--$	$R_3 - R_1$
	$\frac{7}{2}$	1	0	0	$\frac{3}{2}$	1	0	27	27	$\xrightarrow{R_4 + 7R_1}$
	4	-7	0	0	2	0	1	20		

$$\uparrow$$
$$p.c.$$

x	y	z	u	v	w	P	Const.
$-\frac{1}{2}$	1	0	1	$\frac{1}{2}$	0	0	13
0	0	1	1	1	0	0	18
4	0	0	-1	1	1	0	14
$\frac{1}{2}$	0	0	7	$\frac{11}{2}$	0	1	111

The last tableau is in final form. We see that $x = 0$, $y = 13$, $z = 18$, $w = 14$, and $C = -P = -111$.

7. $x = 5/4$, $y = 1/4$, $u = 2$, $v = 3$, and $C = P = 13$.

9. $x = 5$, $y = 10$, $z = 0$, $u = 1$, $v = 2$, and $C = P = 80$.

11. We first write the tableau

x	y	Const.
1	2	4
3	2	6
2	5	

Then obtain the following by interchanging rows and columns:

u	v	Const.
1	3	2
2	2	5
4	6	

From this table we construct the dual problem:
Maximize the objective function
$$P = 4u + 6v \text{ subject to}$$
$$u + 3v \le 2$$
$$2u + 2v \le 5$$
$$u \ge 0,\ v \ge 0$$

Solving the dual problem using the simplex method with x and y as the slack variables, we obtain

	u	v	x	y	P	Const.	Ratio	
p.r. →	1	$\boxed{3}$	1	0	0	2	$\frac{2}{3}$	$\frac{1}{3}R_1$ →
	2	2	0	1	0	5	$\frac{5}{2}$	
	−4	−6	0	0	1	0		

$$\uparrow$$
$$p.c.$$

u	v	x	y	P	Const.	
$\frac{1}{3}$	1	$\frac{1}{3}$	0	0	$\frac{2}{3}$	$R_2 - 2R_1$
2	2	0	1	0	5	$R_3 + 6R_1$ →
−4	−6	0	0	1	0	

	u	v	x	y	P	Const.	Ratio	
p.r. →	$\boxed{\frac{1}{3}}$	1	$\frac{1}{3}$	0	0	$\frac{2}{3}$	2	$3R_1$ →
	$\frac{4}{3}$	0	$-\frac{2}{3}$	1	0	$\frac{11}{3}$	$2\frac{3}{4}$	
	−2	0	2	0	1	4		

$$\uparrow$$
$$p.c.$$

u	v	x	y	P	Const.	
1	3	1	0	0	2	$R_2 - \frac{4}{3}R_1$
$\frac{4}{3}$	0	$-\frac{2}{3}$	1	0	$\frac{11}{3}$	$R_3 + 2R_1$ →
−2	0	2	0	1	4	

u	v	x	y	P	Const.
1	3	1	0	0	2
0	−4	−2	1	0	1
0	6	4	0	1	8

Interpreting the final tableau, we see that $x = 4$, $y = 0$, and $P = C = 8$.

13. We first write the tableau

x	y	Const.
6	1	60
2	1	40
1	1	30
6	4	

Then obtain the following by interchanging rows and columns:

u	v	w	Const.
6	2	1	6
1	1	1	4
60	40	30	

From this table we construct the dual problem:

Maximize $P = 60u + 40v + 30w$ subject to

$$6u + 2v + w \le 6$$

$$u + v + w \le 4$$

$$u \ge 0, \ v \ge 0, \ w \ge 0$$

We solve the problem as follows.

	u	v	w	x	y	P	Const.	Ratio	
p.r. →	$\boxed{6}$	2	1	1	0	0	6	1	$\frac{1}{6}R_1$ →
	1	1	1	0	1	0	4	4	
	−60	−40	−30	0	0	1	0	−−	

↑
p.c

u	v	w	x	y	P	Const.	
1	$\frac{1}{3}$	$\frac{1}{6}$	$\frac{1}{6}$	0	0	1	$\xrightarrow{\begin{array}{c}R_2-R_1\\R_3+60R_1\end{array}}$
1	1	1	0	1	0	4	
−60	−40	−30	0	0	1	0	

	u	v	w	x	y	P	Const.	Ratio	
	1	$\frac{1}{3}$	$\frac{1}{6}$	$\frac{1}{6}$	0	0	1	6	$\xrightarrow{\frac{6}{5}R_2}$
p.r.→	0	$\frac{2}{3}$	$\boxed{\frac{5}{6}}$	$-\frac{1}{6}$	1	0	3	18/5	
	0	−20	−20	10	0	1	60	−−	

$$\underset{\text{p.c.}}{\uparrow}$$

u	v	w	x	y	P	Const.	
1	$\frac{1}{3}$	$\frac{1}{6}$	$\frac{1}{6}$	0	0	1	$\xrightarrow{\begin{array}{c}R_1-\frac{1}{6}R_2\\R_3+20R_2\end{array}}$
0	$\frac{4}{5}$	1	$-\frac{1}{5}$	$\frac{6}{5}$	0	$\frac{18}{5}$	
0	−20	−20	10	0	1	60	

	u	v	w	x	y	P	Const.	Ratio	
p.r.→	1	$\boxed{\frac{1}{5}}$	0	$\frac{1}{5}$	$-\frac{1}{5}$	0	$\frac{2}{5}$	2	$\xrightarrow{5R_1}$
	0	$\frac{4}{5}$	1	$-\frac{1}{5}$	$\frac{6}{5}$	0	$\frac{18}{5}$	9/2	
	0	−4	0	6	24	1	132	−−	

$$\underset{\text{p.c.}}{\uparrow}$$

u	v	w	x	y	P	Const.	
5	1	0	1	−1	0	2	$\xrightarrow{\begin{array}{c}R_2-\frac{4}{5}R_1\\R_3+4R_1\end{array}}$
0	$\frac{4}{5}$	1	$-\frac{1}{5}$	$\frac{6}{5}$	0	$\frac{18}{5}$	
0	−4	0	6	24	1	132	

u	v	w	x	y	P	Const.
5	1	0	1	−1	0	2
−4	0	1	−1	2	0	2
20	0	0	10	20	1	140

The last tableau is in final form. We find that $x = 10$, $y = 20$, and $C = 140$.

15. We first write the tableau

x	y	z	Const.
20	10	1	10
1	1	2	20
200	150	120	

Then obtain the following by interchanging rows and columns:

u	v	Const.
20	1	200
10	1	150
1	2	120
10	20	

From this table we construct the dual problem:

Maximize $P = 10u + 20v$ subject to

$$20u + \quad v \le 200$$
$$10u + \quad v \le 150$$
$$u + 2v \le 120$$
$$u \ge 0, v \ge 0$$

Solving this problem, we obtain the following tableaus:

	u	v	x	y	z	P	Const.	Ratio	
	20	1	1	0	0	0	200	200	
	10	1	0	1	0	0	150	150	$\frac{1}{2}R_3$
p.r. $\rightarrow$	1	$\boxed{2}$	0	0	1	0	120	60	
	-10	-20	0	0	0	1	0		

$$\underset{p.\varsigma}{\uparrow}$$

u	v	x	y	z	P	Const.	
20	1	1	0	0	0	200	$R_1 - R_3$
10	1	0	1	0	0	150	$R_2 - R_3$
$\frac{1}{2}$	1	0	0	$\frac{1}{2}$	0	60	$R_4 + 20R_3$
-10	-20	0	0	0	1	0	

u	v	x	y	z	P	Const.
$\frac{39}{2}$	0	1	0	$-\frac{1}{2}$	0	140
$\frac{19}{2}$	0	0	1	$-\frac{1}{2}$	0	90
$\frac{1}{2}$	1	0	0	$\frac{1}{2}$	0	60
0	0	0	0	10	1	1200

This last tableau is in final form. We find that $x = 0$, $y = 0$, $z = 10$, and $C = 1200$.

17. We first write the tableau

x	y	z	Const.
1	2	2	10
2	1	1	24
1	1	1	16
6	8	4	

Then obtain the following by interchanging rows and columns:

u	v	w	Const.
1	2	1	6
2	1	1	8
2	1	1	4
10	24	16	

From this table we construct the dual problem:
Maximize the objective function
$P = 10u + 24v + 16w$ subject to
$$u + 2v + w \le 6$$
$$2u + v + w \le 8$$
$$2u + v + w \le 4$$
$$u \ge 0, v \ge 0, w \ge 0$$

Solving the dual problem using the simplex method with x, y, and z as slack variables, we obtain

	u	v	w	x	y	z	P	Const.
p.r. →	1	⟨2⟩	1	1	0	0	0	6
	2	1	1	0	1	0	0	8
	2	1	1	0	0	1	0	4
	−10	−24	−16	0	0	0	1	0

Ratio: 3, 8, 4 $\xrightarrow{\frac{1}{2}R_1}$

↑ p.c.

u	v	w	x	y	z	P	Const.
$\frac{1}{2}$	1	$\frac{1}{2}$	$\frac{1}{2}$	0	0	0	3
2	1	1	0	1	0	0	8
2	1	1	0	0	1	0	4
−10	−24	−16	0	0	0	1	0

$\xrightarrow{\begin{array}{l}R_2-R_1\\R_3-R_1\\R_4+24R_1\end{array}}$

	u	v	w	x	y	z	P	Const.
	$\frac{1}{2}$	1	$\frac{1}{2}$	$\frac{1}{2}$	0	0	0	3
	$\frac{3}{2}$	0	$\frac{1}{2}$	$-\frac{1}{2}$	1	0	0	5
p.r. →	$\frac{3}{2}$	0	⟨$\frac{1}{2}$⟩	$-\frac{1}{2}$	0	1	0	1
	2	0	−4	12	0	0	1	72

Ratio: 6, 10, 2 $\xrightarrow{2R_3}$

↑ p.c.

u	v	w	x	y	z	P	Const.
$\frac{1}{2}$	1	$\frac{1}{2}$	$\frac{1}{2}$	0	0	0	3
$\frac{3}{2}$	0	$\frac{1}{2}$	$-\frac{1}{2}$	1	0	0	5
3	0	1	−1	0	2	0	2
2	0	−4	12	0	0	1	72

$\xrightarrow{\begin{array}{l}R_1-\frac{1}{2}R_3\\R_2-\frac{1}{2}R_3\\R_4+4R_3\end{array}}$

u	v	w	x	y	z	P	Const.
−1	1	0	1	0	−1	0	2
0	0	0	0	1	−1	0	4
3	0	1	−1	0	2	0	2
14	0	0	8	0	8	1	80

The solution to the primal problem is $x = 8$, $y = 0$, $z = 8$, and $C = 80$.

19.   We first write

x	y	z	Const.
2	4	3	6
6	0	1	2
0	6	2	4
30	12	20	

Then obtain the following by interchanging rows and columns:

u	v	w	Const.
2	6	0	30
4	0	6	12
3	1	2	20
6	2	4	

From this table we construct the dual problem:

Maximize $P = 6u + 2v + 4w$ subject to

$$2u + 6v \qquad \leq 30$$
$$4u + \qquad 6w \leq 12$$
$$3u + v + 2w \leq 20$$
$$u \geq 0,\ v \geq 0,\ w \geq 0$$

Using the simplex method, we obtain

	u	v	w	x	y	z	P	Const.	Ratio	
	2	6	0	1	0	0	0	30	15	
$p.r. \rightarrow$	$\boxed{4}$	0	6	0	1	0	0	12	3	$\xrightarrow{\frac{1}{4}R_2}$
	3	1	2	0	0	1	0	20	$\frac{20}{3}$	
	−6	−2	−4	0	0	0	1	0		

$\uparrow$
$p.c.$

u	v	w	x	y	z	P	Const.
2	6	0	1	0	0	0	30
1	0	$\frac{3}{2}$	0	$\frac{1}{4}$	0	0	3
3	1	2	0	0	1	0	20
-6	-2	-4	0	0	0	1	0

$$\xrightarrow{\begin{array}{c} R_1-2R_2 \\ R_3-3R_2 \\ R_4+6R_2 \end{array}}$$

	u	v	w	x	y	z	P	Const.	Ratio
$p.r. \rightarrow$	0	$\boxed{6}$	-3	1	$-\frac{1}{2}$	0	0	24	4
	1	0	$\frac{3}{2}$	0	$\frac{1}{4}$	0	0	3	---
	0	1	$-\frac{5}{2}$	0	$-\frac{3}{4}$	1	0	11	11
	0	-2	5	0	$\frac{3}{2}$	0	1	18	

$p.c.$ (under v)

$$\xrightarrow{\frac{1}{6}R_1}$$

u	v	w	x	y	z	P	Const.
0	1	$-\frac{1}{2}$	$\frac{1}{6}$	$-\frac{1}{12}$	0	0	4
1	0	$\frac{3}{2}$	0	$\frac{1}{4}$	0	0	3
0	1	$-\frac{5}{2}$	0	$-\frac{3}{4}$	1	0	11
0	-2	5	0	$\frac{3}{2}$	0	1	18

$$\xrightarrow{\begin{array}{c} R_3-R_1 \\ R_4+2R_1 \end{array}}$$

u	v	w	x	y	z	P	Const.
0	1	$-\frac{1}{2}$	$\frac{1}{6}$	$-\frac{1}{12}$	0	0	4
1	0	$\frac{3}{2}$	0	$\frac{1}{4}$	0	0	3
0	0	-2	$-\frac{1}{6}$	$-\frac{2}{3}$	1	0	7
0	0	4	$\frac{1}{3}$	$\frac{4}{3}$	0	1	26

The last tableau is in final form. We find $x = 1/3$, $y = 4/3$, $z = 0$, and $C = 26$.

21. Let x denote the number of type-A vessels and y the number of type-B vessels to be operated. Then the problem is

Maximize $C = 44,000x + 54,000y$ subject to

$$60x + 80y \geq 360$$

$$160x + 120y \geq 680$$

$$x \geq 0, y \geq 0$$

We first write down the following tableau for the primal problem.

x	y	Constant
60	80	360
160	120	680
44,000	54,000	

Next, we interchange the columns and rows of the frequency tableaus obtaining

u	v	Constant
60	160	44,000
80	120	54,000
360	680	

Proceeding, we are led to the dual problem

Maximize $P = 360u + 680v$ subject to

$$60u + 160v \leq 44000$$

$$80u + 120v \leq 54000$$

$$u \geq 0, v \geq 0$$

Let x and y be slack variables. We obtain the following tableaus:

	u	v	x	y	P	Const.	Ratio	
p.r. $\rightarrow$	60	160	1	0	0	44,000	275	$\frac{1}{160}R_1$
	80	120	0	1	0	54,000	450	
	−360	−680	0	0	1	0		

p.c.

u	v	x	y	P	Const.	
$\frac{3}{8}$	1	$\frac{1}{160}$	0	0	275	$R_2 - 120R_1$
80	120	0	1	0	54,000	$R_3 + 680R_1$
−360	−680	0	0	1	0	

	u	v	x	y	P	Const.		Ratio
	$\frac{3}{8}$	1	$\frac{1}{160}$	0	0	275		$733\frac{1}{3}$
p.r. →	$\boxed{35}$	0	$-\frac{3}{4}$	1	0	21000		600
	-105	0	$\frac{17}{4}$	0	1	187000		

$\xrightarrow{\frac{1}{35}R_1}$

p.c.

u	v	x	y	P	Const.
$\frac{3}{8}$	1	$\frac{1}{160}$	0	0	275
$\boxed{1}$	0	$-\frac{3}{140}$	$\frac{1}{35}$	0	600
-105	0	$\frac{17}{4}$	0	1	187000

$\xrightarrow[R_3+105R_2]{R_1-\frac{3}{8}R_2}$

u	v	x	y	P	Const.
0	1	$\frac{1}{70}$	$-\frac{3}{280}$	0	50
1	0	$-\frac{3}{140}$	$\frac{1}{35}$	0	600
0	0	2	3	1	250000

The last tableau is in final form. The fundamental theorem of duality tells us that the solution to the primal problem is $x = 2$, $y = 3$ with a minimum value for C of 250,000. Thus, Deluxe River Cruises should use 2 type-A vessels and 3 type-B vessels. The minimum operating cost is $250,000.

23. Let x and y denote, respectively, the number of advertisements to be placed in newspaper I and newspaper II. Then the problem is

Minimize $C = 1000x + 800y$ subject to
$$70000x + 10000y \geq 2000000$$
$$40000x + 20000y \geq 1400000$$
$$20000x + 40000y \geq 1000000$$
$$x \geq 0, y \geq 0$$

or upon simplification:

Minimize $C = 1000x + 800y$ subject to
$$7x + \ y \geq 200$$
$$2x + \ y \geq 70$$
$$x + 2y \geq 50$$
$$x \geq 0, y \geq 0$$

We first write down the following tableau for the primal problem

	x	y	Const.
	7	1	200
	2	1	70
	1	2	50
	1000	800	

Next, we interchange the columns and rows of the foregoing tableau:

u	v	w	Const.
7	2	1	1000
1	1	2	800
200	70	50	

Proceeding, we obtain the following dual problem
Maximize $P = 200u + 70v + 50w$ subject to
$$7u + 2v + w \le 1000$$
$$u + v + 2w \le 800$$
$$u \ge 0, v \ge 0, w \ge 0$$

Let x and y denote the slack variables. We obtain the following tableaus:

	u	v	w	x	y	P	Const.	Ratio
p.r. →	$\boxed{7}$	2	1	1	0	0	1000	$142\frac{6}{7}$
	1	1	2	0	1	0	800	800
	−200	−70	−50	0	0	1	0	

$\uparrow$
p.c.

$\xrightarrow{\frac{1}{7}R_1}$

u	v	w	x	y	P	Const.
$\boxed{1}$	$\frac{2}{7}$	$\frac{1}{7}$	$\frac{1}{7}$	0	0	$\frac{1000}{7}$
1	1	2	0	1	0	800
−200	−70	−50	0	0	1	0

$\xrightarrow[R_3 + 200R_1]{R_2 - R_1}$

u	v	w	x	y	P	Const.	Ratio
1	$\frac{2}{7}$	$\frac{1}{7}$	$\frac{1}{7}$	0	0	$\frac{1000}{7}$	1000
p.r. → 0	$\frac{5}{7}$	$\boxed{\frac{13}{7}}$	$-\frac{1}{7}$	1	0	$\frac{4600}{7}$	$657\frac{1}{7}$
0	$-\frac{90}{7}$	$-\frac{150}{7}$	$\frac{200}{7}$	0	1	$\frac{200000}{7}$	

$\xrightarrow{\frac{7}{13}R_2}$

p.c. (under w column)

u	v	w	x	y	P	Const.
1	$\frac{2}{7}$	$\frac{1}{7}$	$\frac{1}{7}$	0	0	$\frac{1000}{7}$
0	$\frac{5}{13}$	$\boxed{1}$	$-\frac{1}{13}$	$\frac{7}{13}$	0	$\frac{4600}{7}$
0	$-\frac{90}{7}$	$-\frac{150}{7}$	$\frac{200}{7}$	0	1	$\frac{200000}{7}$

$\xrightarrow[R_1 + \frac{150}{7}R_2]{R_1 - \frac{1}{7}R_2}$

u	v	w	x	y	P	Const.	Ratio
p.r. → 1	$\boxed{\frac{3}{13}}$	0	$\frac{2}{13}$	$-\frac{1}{13}$	0	$\frac{1200}{13}$	400
0	$\frac{5}{13}$	1	$-\frac{1}{13}$	$\frac{7}{13}$	0	$\frac{4600}{13}$	920
0	$-\frac{60}{13}$	0	$\frac{300}{13}$	$\frac{150}{13}$	1	$\frac{470000}{13}$	

$\xrightarrow{\frac{13}{3}R_1}$

p.c. (under v column)

u	v	w	x	y	P	Const.
$\frac{13}{3}$	$\boxed{1}$	0	$\frac{2}{3}$	$-\frac{1}{3}$	0	400
0	$\frac{5}{13}$	1	$-\frac{1}{13}$	$\frac{7}{13}$	0	$\frac{4600}{13}$
0	$-\frac{60}{13}$	0	$\frac{300}{13}$	$\frac{150}{13}$	1	$\frac{470000}{13}$

$\xrightarrow[R_3 + \frac{60}{13}R_1]{R_2 - \frac{5}{13}R_1}$

u	v	w	x	y	P	Const.
$\frac{13}{3}$	1	0	$\frac{2}{3}$	$-\frac{1}{3}$	0	400
$-\frac{5}{2}$	0	1	$-\frac{1}{3}$	$\frac{2}{3}$	0	200
20	0	0	30	10	1	38000

The last tableau is in final form. The fundamental theorem of duality tells us that the solution to the primal problem is $x = 30$, $y = 10$, and $C = 38000$. Therefore, Everest Deluxe World Travel should place 30 advertisements in newspaper I and 10 advertisements in newspaper II at a minimum cost of $38,000.

25. The given data may be summarized as follows:

	Orange Juice	Grapefruit Juice
Vitamin A	60 I.U.	120 I.U.
Vitamin C	16 I.U.	12 I.U.
Calories	14	11

Suppose x ounces of orange juice and y ounces of pink-grapefruit juice are required for each glass of the blend. Then the problem is

Minimize $C = 14x + 11y$ subject to

$$60x + 120y \geq 1200$$
$$16x + 12y \geq 200$$
$$x \geq 0, y \geq 0$$

To construct the dual problem, we first write down the tableau

x	y	Const.
60	120	1200
16	12	200
14	11	

Then obtain the following by interchanging rows and columns:

u	v	Const.
60	16	14
120	12	11
1200	200	

From this table we construct the dual problem:

Maximize $P = 1200u + 200v$ subject to

$$60u + 16v \leq 14$$
$$120u + 12v \leq 11$$
$$u \geq 0, v \geq 0$$

The initial tableau is

u	v	x	y	P	Const.
60	16	1	0	0	14
120	12	0	1	0	11
−1200	−200	0	0	1	0

Using the following sequence of row operations,

1. $\frac{1}{120} R_2$ 2. $R_1 - 60R_2$, $R_3 + 1200R_2$ 3. $\frac{1}{10} R_1$ 4. $R_2 - \frac{1}{10} R_1$, $R_3 + 80R_1$

we obtain the final tableau

u	v	x	y	P	Const.
0	1	$\frac{1}{10}$	$-\frac{1}{20}$	0	$\frac{17}{20}$
0	0	$-\frac{1}{100}$	$\frac{1}{75}$	0	$\frac{1}{150}$
0	0	8	6	1	178

We conclude that the owner should use 8 ounces of orange juice and 6 ounces of grapefruit juice per glass of the blend for a minimal calorie count of 178.

27.   True. To maximize P, one maximizes $-C$. Since the minimization problem has a unique solution, the negative of that solution is the solution of the maximization problem.

USING TECHNOLOGY EXERCISES 4.2, page 264

1. $x = \frac{4}{3}$, $y = \frac{10}{3}$, $z = 0$, and $C = \frac{14}{3}$

3. $x = 0.9524$, $y = 4.2857$, $z = 0$, and $C = 6.0952$.

5.   a. $x = 3, y = 2$, and $P = 17$ b. $\frac{8}{3} \le c_1 \le 8; \frac{3}{2} \le c_2 \le \frac{9}{2}$
      c. $8 \le b_1 \le 24; 4 \le b_2 \le 12$ d. $\frac{5}{4}; \frac{1}{4}$ e. Both constraints are binding

7.   a. $x = 4, y = 0$, and $C = 18$ b. $0 \le c_1 \le \frac{5}{2}; 4 \le c_2 < \infty$
      c. $3 \le b_1 < \infty; -\infty < b_2 \le 4$ d. 2;0   e. Both constraints are binding.

EXERCISES 4.3, page 275

1.   Maximize $P = -C = -2x + 3y$ subject to
$$-3x - 5y \le -20$$
$$3x + y \le 16$$
$$-2x + y \le 1$$
$$x \ge 0, y \ge 0$$

3.   Maximize $P = -C = -5x - 10y - z$ subject to

$$-2x - y - z \le -4$$
$$-x - 2y - 2z \le -2$$
$$2x + 4y + 3z \le 12$$
$$x \ge 0,\ y \ge 0,\ \text{and } z \ge 0$$

5. We set up the tableau and solve the problem using the simplex method:

x	y	u	v	P	Const	Ratio
2	[5]	1	0	0	20	4
1	−5	0	1	0	−5	1
−1	−2	0	0	1	0	

p.r. → second row, p.c. ↑ under y

$\xrightarrow{-\frac{1}{5}R_2}$

x	y	u	v	P	Const
2	5	1	0	0	20
$-\frac{1}{5}$	1	0	$-\frac{1}{5}$	0	1
−1	−2	0	0	1	0

$\xrightarrow[R_3+2R_2]{R_1-5R_2}$

x	y	u	v	P	Const	Ratio
[3]	0	1	1	0	15	5
$-\frac{1}{5}$	1	0	$-\frac{1}{5}$	0	1	--
$-\frac{7}{5}$	0	0	$-\frac{2}{5}$	1	2	

$\xrightarrow{\frac{1}{3}R_1}$

x	y	u	v	P	Const
1	0	$\frac{1}{3}$	$\frac{1}{3}$	0	5
$-\frac{1}{5}$	1	0	$-\frac{1}{5}$	0	1
$-\frac{7}{5}$	0	0	$-\frac{2}{5}$	1	2

$\xrightarrow[R_3+\frac{7}{5}R_1]{R_2+\frac{1}{5}R_1}$

x	y	u	v	P	Const
1	0	$\frac{1}{3}$	$\frac{1}{3}$	0	5
0	1	$\frac{1}{15}$	$-\frac{2}{15}$	0	2
0	0	$\frac{7}{15}$	$\frac{1}{15}$	1	9

The maximum value of P is 9 when $x = 5$ and $y = 2$.

7. We first rewrite the problem as a maximization problem with inequality constraints using $\le$, obtaining the following equivalent problem:

$$\text{Maximize } P = -C = 2x - y \text{ subject to}$$
$$x + 2y \le 6$$
$$3x + 2y \le 12$$
$$x \ge 0,\ y \ge 0$$

Following the procedure outlined for nonstandard problems, we have

x	y	u	v	P	Const		Ratio	
1	2	1	0	0	6		6	
3	2	0	1	0	12		4	$\frac{1}{3}R_2$
−2	1	0	0	1	0			

p.r. → (row with boxed 3)

p.c. ↑ (x column)

x	y	u	v	P	Const
1	2	1	0	0	6
1	$\frac{2}{3}$	0	$\frac{1}{3}$	0	4
−2	1	0	0	1	0

$\xrightarrow[R_3+2R_2]{R_1-R_2}$

x	y	u	v	P	Const
0	$\frac{4}{3}$	1	$-\frac{1}{3}$	0	2
1	$\frac{2}{3}$	0	$\frac{1}{3}$	0	4
0	$\frac{7}{3}$	0	$\frac{2}{3}$	1	8

We conclude that C attains a minimum value of -8 when $x = 4$ and $y = 0$.

9. Using the simplex method we have

x	y	u	v	P	Const		Ratio	
1	3	1	0	0	6		6	
−2	3	0	1	0	−6		3	$-\frac{1}{2}R_2$
−1	−4	0	0	1	0			

p.r. → (row with boxed −2)

p.c. ↑ (x column)

x	y	u	v	P	Const
1	3	1	0	0	6
1	$-\frac{3}{2}$	0	$-\frac{1}{2}$	0	3
−1	−4	0	0	1	0

$\xrightarrow[R_3+R_2]{R_1-R_2}$

x	y	u	v	P	Const
0	$\frac{9}{2}$	1	$\frac{1}{2}$	0	3
1	$-\frac{3}{2}$	0	$-\frac{1}{2}$	0	3
0	$-\frac{11}{2}$	0	$-\frac{1}{2}$	1	3

$\xrightarrow{\frac{2}{9}R_1}$

x	y	u	v	P	Const
0	1	$\frac{2}{9}$	$\frac{1}{9}$	0	$\frac{2}{3}$
1	$-\frac{3}{2}$	0	$-\frac{1}{2}$	0	3
0	$-\frac{11}{2}$	0	$-\frac{1}{2}$	1	3

$\xrightarrow[R_3+\frac{11}{2}R_1]{R_2+\frac{3}{2}R_1}$

x	y	u	v	P	Const
0	1	$\frac{2}{9}$	$\frac{1}{9}$	0	$\frac{2}{3}$
1	0	$\frac{1}{3}$	$-\frac{1}{3}$	0	4
0	0	$\frac{11}{9}$	$\frac{1}{9}$	1	$\frac{20}{3}$

We conclude that P attains a maximum value of 20/3, when $x = 4$ and $y = 2/3$.

11. We rewrite the problem as

$$\text{Maximize } P = x + 2y \text{ subject to}$$
$$2x + 3y \le 12$$
$$-x + 3y \le 3$$
$$-x + 3y \ge 3$$
$$x \ge 0,\ y \ge 0$$

The initial tableau is

x	y	u	v	w	P	Const.
2	3	1	0	0	0	12
−1	3	0	1	0	0	3
1	−3	0	0	1	0	−3
−1	−2	0	0	0	1	0

Using the following sequence of row operations

1. $-\frac{1}{3}R_3$ 2. $R_1 - 3R_3,\ R_2 - 3R_3,\ R_4 + 2R_3$ 3. $\frac{1}{3}R_1$

4. $R_3 + \frac{1}{3}R_1,\ R_4 + \frac{5}{3}R_1$ 5. $R_1 - \frac{1}{3}R_2,\ R_3 + \frac{2}{9}R_2,\ R_4 + \frac{1}{9}R_2$

we obtain the final tableau

x	y	u	v	w	P	Const.
1	0	$\frac{1}{3}$	$-\frac{1}{3}$	0	0	3
0	0	0	1	1	0	0
0	1	$\frac{1}{9}$	$\frac{2}{9}$	0	0	2
0	0	$\frac{5}{9}$	$\frac{1}{9}$	0	1	7

We conclude that P attains a maximum value of 7 when $x = 3$ and $y = 2$.

13. We rewrite the problem as

$$\text{Maximize } P = 5x + 4y + 2z \text{ subject to}$$
$$x + 2y + 3z \le 24$$
$$-x + y - z \le -6$$
$$x \ge 0,\ y \ge 0,\ z \ge 0$$

The initial tableau is

x	y	z	u	v	P	Const.
1	2	3	1	0	0	24
−1	1	−1	0	1	0	−6
−5	−4	−2	0	0	1	0

Using the following sequence of row operations
1. $-R_2$ 2. $R_1 - 2R_2$, $R_3 + 5R_2$ 3. $\frac{1}{3}R_1$ 4. $R_2 + R_1$, $R_3 + 9R_1$
5. $3R_1$ 6, $R_2 + \frac{2}{3}R_1$; $R_3 + 2R_1$

we obtain the final tableau

x	y	z	u	v	P	Const.
0	3	2	1	1	0	18
1	2	3	1	0	0	24
0	6	13	5	0	1	120

We deduce that P attains a maximum value of 120 when $x = 24$, $y = 0$, and $z = 0$.

15. The problem is to maximize $P = -C = -x + 2y - z$ subject to the given constraints.
The initial tableau is

x	y	z	u	v	w	P	Const.
1	−2	3	1	0	0	0	10
2	1	−2	0	1	0	0	15
2	1	3	0	0	1	0	20
1	−2	1	0	0	0	1	0

Using the following sequence of row operations,
1. $R_1 + 2R_2$, $R_3 - R_2$, $R_4 + 2R_2$ 2. $\frac{1}{5}R_3$ 3. $R_1 + R_3$, $R_2 + 2R_3$, $R_4 + 3R_3$

we obtain the final tableau

x	y	z	u	v	w	P	Const.
5	0	0	1	$\frac{9}{5}$	$\frac{1}{5}$	0	41
2	1	0	0	$\frac{3}{5}$	$\frac{2}{5}$	0	17
0	0	1	0	$-\frac{1}{5}$	$\frac{1}{5}$	0	1
5	0	0	0	$\frac{7}{5}$	$\frac{3}{5}$	1	33

We conclude that C attains a minimum value of -33 when $x = 0$, $y = 17$, $z = 1$, and
$C = -P = -33$.

17. Rewriting the third constraint as $-x + 2y - z \leq -4$, we obtain the following initial
tableau

x	y	z	u	v	w	P	Const.
1	2	3	1	0	0	0	28
2	3	-1	0	1	0	0	6
-1	2	-1	0	0	1	0	-4
-2	-1	-1	0	0	0	1	0

Using the following sequence of row operations,

1. $\frac{1}{2}R_2$ 2. $R_1 - R_2$, $R_3 + R_2$, $R_4 + 2R_2$ 3. $-\frac{2}{3}R_3$
4. $R_1 - \frac{7}{2}R_3$, $R_2 + \frac{1}{2}R_3$, $R_4 + 2R_3$ 5. $\frac{3}{26}R_1$ 6. $R_2 - \frac{1}{3}R_1$; $R_3 + \frac{7}{3}R_1$, $R_4 + \frac{8}{3}R_1$
7. $\frac{26}{7}R_1$ 8. $R_2 + \frac{11}{26}R_1$, $R_3 + \frac{1}{26}R_1$, $R_4 + \frac{8}{13}R_1$

we obtain the final tableau

x	y	z	u	v	w	P	Const.
0	$\frac{26}{7}$	0	$\frac{3}{7}$	$\frac{2}{7}$	1	0	$\frac{68}{7}$
1	$\frac{11}{7}$	0	$\frac{1}{7}$	$\frac{3}{7}$	1	0	$\frac{46}{7}$
0	$\frac{1}{7}$	1	$\frac{2}{7}$	$-\frac{1}{7}$	0	0	$\frac{50}{7}$
0	$\frac{16}{7}$	0	$\frac{4}{7}$	$\frac{5}{7}$	0	1	$\frac{142}{7}$

from which we deduce that P attains a maximum value of 142/7 when $x = 46/7$, $y = 0$, and $z = 50/7$.

19. Rewriting the third constraint $(2x + y + z = 10)$ in the form
$$2x + y + z \geq 10 \quad \text{and} \quad -2x - y - z \leq -10,$$
we obtain the following initial tableau.

x	y	z	t	u	v	w	P	Const.
1	2	1	1	0	0	0	0	20
3	1	0	0	1	0	0	0	30
2	1	1	0	0	1	0	0	10
-2	-1	-1	0	0	0	1	0	-10
-1	-2	-3	0	0	0	0	1	0

Using the following sequence of row operations
1. $-R_4$ 2. $R_1 - R_4$, $R_3 - R_4$, $R_5 + 3R_4$ 3. $R_1 - R_3$, $R_4 + R_3$, $R_5 + 3R_3$
we obtain the final tableau

x	y	z	t	u	v	w	P	Const.
−1	1	0	1	0	−1	0	0	10
3	1	0	0	1	0	0	0	30
0	0	0	0	0	1	1	0	0
2	1	1	0	0	1	0	0	10
5	1	0	0	0	3	0	1	30

We conclude that P attains a maximum value of 30 when $x = 0$, $y = 0$, and $z = 10$.

21. Let x and y denote the number of acres of crops A and B, respectively to be planted. Then the problem is

Maximize $P = 150x + 200y$ subject to the constraints

$$x + \quad y \leq 150$$
$$40x + 60y \leq 7400$$
$$20x + 25y \leq 3300$$
$$x \geq 80$$
$$x \geq 0, y \geq 0$$

Using the simplex method, we obtain

x	y	u	v	w	z	P	Const.		Ratio	
1	1	1	0	0	0	0	150		150	
40	60	0	1	0	0	0	7400		185	
20	25	0	0	1	0	0	3300		165	$\xrightarrow{-R_4}$
−1	0	0	0	0	1	0	−80		−−	
−150	−200	0	0	0	0	1	0			

x	y	u	v	w	z	P	Const.	
1	1	1	0	0	0	0	150	$R_1 - R_4$
40	60	0	1	0	0	0	7400	$R_2 - 40R_4$
20	25	0	0	1	0	0	3300	$R_3 - 20R_4$
1	0	0	0	0	−1	0	80	$\xrightarrow{R_5 + 150R_4}$
−150	−200	0	0	0	0	1	0	

x	y	u	v	w	z	P	Const.
0	1	1	0	0	1	0	70
0	60	0	1	0	40	0	4200
0	25	0	0	1	20	0	1700
1	0	0	0	0	-1	0	80
0	-200	0	0	0	-150	1	12,000

$$\xrightarrow{\frac{1}{25}R_3}$$

x	y	u	v	w	z	P	Const.
0	1	1	0	0	1	0	70
0	60	0	1	0	40	0	4200
0	1	0	0	$\frac{1}{25}$	$\frac{4}{5}$	0	68
1	0	0	0	0	-1	0	80
0	-200	0	0	0	-150	1	12,000

$$\xrightarrow[\substack{R_2-60R_3 \\ R_5+200R_3}]{R_1-R_3}$$

x	y	u	v	w	z	P	Const.
0	0	1	0	$-\frac{1}{25}$	$\frac{1}{5}$	0	2
0	0	0	1	$-\frac{12}{5}$	-8	0	120
0	1	0	0	$\frac{1}{25}$	$\frac{4}{5}$	0	68
1	0	0	0	0	-1	0	80
0	0	0	0	8	10	1	25,600

We conclude that the farmer should plant 80 acres of crop A and 68 acres of crop B to realize a maximum profit of $25,600.

23. Let x and y denote the amount (in dollars) invested in home loans and commercial-development loans, respectively. Then the problem is

Maximize $P = 0.08x + 0.06y$ subject to
$$-x + 3y \leq 0$$
$$y \geq 10,000,000$$
$$x + y = 60,000,000$$
$$x \geq 0, \ y \geq 0$$

Substituting $x = 60,000,000 - y$ into the first equation and the first and second inequalities, we have

Maximize $P = 0.08(60{,}000{,}000 - y) + 0.06y$
$= 4{,}800{,}000 - 0.02y$ subject to
$$y \le 15{,}000{,}000$$
$$y \ge 10{,}000{,}000$$
$$x \ge 0, y \ge 0.$$

Using the simplex method, we have

	y	u	v	P	Const.	
	1	1	0	0	15,000,000	$\xrightarrow{-R_2}$
$p.r.\rightarrow$	-1	0	1	0	$-10{,}000{,}000$	
	0.02	0	0	1	4,800,000	

$$\underset{p.c.}{\uparrow}$$

y	u	v	P	Const.
0	1	1	0	5,000,000
1	0	-1	0	10,000,000
0	0	0.02	1	4,600,000

We conclude that the bank should extend $50 million in home loans, $10 million of commercial-development loans to attain a maximum return of $4.6 million.

25. Let x, y, and z denote the number of units of products A, B, and C manufactured by the company. Then the linear programming problem is

Maximize $P = 18z + 12y + 15z$ subject to
$$2x + y + 2z \le 900$$
$$3x + y + 2z \le 1080$$
$$2x + 2y + z \le 840$$
$$x - y + z \le 0$$
$$x \ge 0, y \ge 0, z \ge 0$$

The initial tableau is

x	y	z	t	u	v	w	P	Const.
2	1	2	1	0	0	0	0	900
3	1	2	0	1	0	0	0	1080
2	2	1	0	0	1	0	0	840
1	−1	1	0	0	0	1	0	0
−18	−12	−15	0	0	0	0	1	0

Using the following sequence of row operations,

1. $R_1 - 2R_4$, $R_2 - 3R_4$, $R_3 - 2R_4$, $R_5 + 18R_4$ 2. $\frac{1}{4}R_3$

3. $R_1 - 3R_3$, $R_2 - 4R_3$, $R_4 + R_3$, $R_5 + 30R_3$ 4. $\frac{4}{3}R_4$

5. $R_1 - \frac{3}{4}R_4$, $R_3 + \frac{1}{4}R_4$, $R_5 + \frac{9}{2}R_4$

we obtain the final tableau

x	y	z	t	u	v	w	P	Const.
−1	0	0	1	0	−1	−1	0	60
0	0	0	0	1	−1	−1	0	240
$\frac{1}{3}$	1	0	0	0	$\frac{1}{3}$	$-\frac{1}{3}$	0	280
$\frac{4}{3}$	0	1	0	0	$\frac{1}{3}$	$\frac{2}{3}$	0	280
−6	0	0	0	0	9	6	1	7560

and conclude that the company should produce 0 units of product A, 280 units of product B, and 280 units of product C to realize a maximum profit of \$7,560.

27. Let x denote the number of ounces of food A and y denote the number of ounces of food B used in the meal. Then the problem is to minimize the amount of cholesterol in the meal. Thus, the linear programming problem is

Maximize $P = -C = -2x - 5y$ subject to
$$30x + 25y \geq 400$$
$$x + \tfrac{1}{2}y \geq 10$$
$$2x + 5y \geq 40$$
$$x \geq 0, y \geq 0$$

The initial tableau is

x	y	u	v	w	C	Const.
−30	−25	1	0	0	0	−400
−1	$-\frac{1}{2}$	0	1	0	0	−10
−2	−5	0	0	1	0	−40
2	5	0	0	0	1	0

Using the following sequence of row operations

1. $-R_2$ 2. $R_1 + 30R_2$; $R_3 + 2R_2$; $R_4 - 2R_2$ 3. $-\frac{1}{4}R_3$

4. $R_1 + 10R_2$; $R_2 - \frac{1}{2}R_3$; $R_4 - 4R_3$ 5. $-\frac{1}{25}R_1$ 6. $R_2 + \frac{5}{4}R_1$; $R_3 - \frac{1}{2}R_1$

we obtain the final tableau

x	y	u	v	w	C	Const.
0	0	$-\frac{1}{25}$	1	$\frac{1}{10}$	0	2
1	0	$-\frac{1}{20}$	1	$\frac{1}{4}$	0	10
0	1	$\frac{1}{50}$	0	$-\frac{3}{10}$	0	4
0	0	0	0	1	1	−40

Thus, the minimum content of cholesterol is 40mg when 10 ounces of food A and 4 ounces of food B are used.(Since the u-column is not in unit form, we see that the problem has multiple solutions.)

CHAPTER 4 REVIEW EXERCISES, page 279

1. This is a regular linear programming problem. Using the simplex method with u and v as slack variables, we obtain the following sequence of tableaus:

	x	y	u	v	P	Const
p.r. →	1	3	1	0	0	15
	4	1	0	1	0	16
	−3	−4	0	0	1	0

Ratio: 5, 16 $\xrightarrow{\frac{1}{3}R_1}$

x	y	u	v	P	Const
$\frac{1}{3}$	1	$\frac{1}{3}$	0	0	5
4	1	0	1	0	16
−3	−4	0	0	1	0

$\xrightarrow{\substack{R_2 - R_1 \\ R_3 + 4R_1}}$

p.c.

	x	y	u	v	P	Const.		Ratio
	$\frac{1}{3}$	1	$\frac{1}{3}$	0	0	5		15
p.c. →	$\boxed{\frac{11}{3}}$	0	$-\frac{1}{3}$	1	0	11	$\xrightarrow{\frac{3}{11}R_2}$	3
	$-\frac{5}{3}$	0	$\frac{4}{3}$	0	1	20		

↑
p.c.

x	y	u	v	P	Const.
$\frac{1}{3}$	1	$\frac{1}{3}$	0	0	5
1	0	$-\frac{1}{11}$	$\frac{3}{11}$	0	3
$-\frac{5}{3}$	0	$\frac{4}{3}$	0	1	20

$\xrightarrow{\substack{R_1-\frac{1}{3}R_2 \\ R_3+\frac{5}{3}R_2}}$

x	y	u	v	P	Const.
0	1	$\frac{4}{11}$	$-\frac{1}{11}$	0	4
1	0	$-\frac{1}{11}$	$\frac{3}{11}$	0	3
0	0	$\frac{13}{11}$	$\frac{5}{11}$	1	25

and conclude that $x = 3$, $y = 4$, $u = 0$, $v = 0$, and $P = 25$.

3. Using the simplex method to solve this regular linear programming problem we have

	x	y	z	u	v	P	Const.		Ratio
p.r. →	1	2	$\boxed{3}$	1	0	0	12		4
	1	-3	2	0	1	0	10	$\xrightarrow{\frac{1}{3}R_1}$	5
	-2	-3	-5	0	0	1	0		

↑
p.c.

x	y	z	u	v	P	Const.
$\frac{1}{3}$	$\frac{2}{3}$	1	$\frac{1}{3}$	0	0	4
1	-3	2	0	1	0	10
-2	-3	-5	0	0	1	0

$\xrightarrow{\substack{R_2-2R_1 \\ R_3+5R_1}}$

	x	y	z	u	v	P	Const.		Ratio
	$\frac{1}{3}$	$\frac{2}{3}$	1	$\frac{1}{3}$	0	0	4		12
p.r. →	$\boxed{\frac{1}{3}}$	$-\frac{13}{3}$	0	$-\frac{2}{3}$	1	0	2	$\xrightarrow{3R_2}$	6
	$-\frac{1}{3}$	$\frac{1}{3}$	0	$\frac{5}{3}$	0	1	20		

↑
p.c.

	x	y	z	u	v	P	Const.	
	$\frac{1}{3}$	$\frac{2}{3}$	1	$\frac{1}{3}$	0	0	4	$R_1-\frac{1}{3}R_2$
	1	−13	0	−2	3	0	6	$R_3+\frac{1}{3}R_2$
	$-\frac{1}{3}$	$\frac{1}{3}$	0	$\frac{5}{3}$	0	1	20	$\longrightarrow$

x	y	z	u	v	P	Const.
0	1	$\frac{1}{5}$	$\frac{1}{5}$	$-\frac{1}{5}$	0	$\frac{2}{5}$
0	−13	0	−2	3	0	6
0	−4	0	1	1	1	22

$$\begin{array}{c} R_2+13R_1 \\ R_3+4R_1 \\ \longrightarrow \end{array}$$

x	y	z	u	v	P	Const.
0	1	$\frac{1}{5}$	$\frac{1}{5}$	$-\frac{1}{5}$	0	$\frac{2}{5}$
1	0	$\frac{13}{5}$	$\frac{3}{5}$	$\frac{2}{5}$	0	$\frac{56}{5}$
0	0	$\frac{4}{5}$	$\frac{9}{5}$	$\frac{1}{5}$	1	$\frac{118}{5}$

We conclude that the P attains a maximum value of 23.6 when $x = 11.2$, $y = 0.4$, $z = 0$, $u = 0$, and $v = 0$.

5.   We first write the tableau

x	y	Const.
2	3	6
2	1	4
3	2	

Then obtain the following by interchanging rows and columns:

u	v	Const.
2	2	3
3	1	2
6	4	

From this table we construct the dual problem:
   Maximize the objective function $P = 6u + 4v$ subject to the constraints
$$2u + 2v \le 3$$
$$3u + v \le 2$$
$$u \ge 0, \; v \ge 0$$

Using the simplex method, we have

	u	v	x	y	P	Const
	2	2	1	0	0	3
p.r. →	[3]	1	0	1	0	2
	-6	-4	0	0	1	0

Ratio
3/2
2/3

$\xrightarrow{\frac{1}{3}R_2}$

↑
p.c.

u	v	x	y	P	Const
2	2	1	0	0	3
1	$\frac{1}{3}$	0	$\frac{1}{3}$	0	$\frac{2}{3}$
-6	-4	0	0	1	0

$\xrightarrow[R_3+6R_2]{R_1-2R_2}$

u	v	x	y	P	Const
0	$\boxed{\frac{4}{3}}$	1	$-\frac{2}{3}$	0	$\frac{5}{3}$
1	$\frac{1}{3}$	0	$\frac{1}{3}$	0	$\frac{2}{3}$
0	-2	0	2	1	4

Ratio
5/4
2

$\xrightarrow{\frac{3}{4}R_1}$

u	v	x	y	P	Const
0	1	$\frac{3}{4}$	$-\frac{1}{2}$	0	$\frac{5}{4}$
1	$\frac{1}{3}$	0	$\frac{1}{3}$	0	$\frac{2}{3}$
0	-2	0	2	1	4

$\xrightarrow[R_3+2R_1]{R_2-\frac{1}{3}R_1}$

u	v	x	y	P	Const
0	1	$\frac{3}{4}$	$-\frac{1}{2}$	0	$\frac{5}{4}$
1	0	$-\frac{1}{4}$	$\frac{1}{2}$	0	$\frac{1}{4}$
0	0	$\frac{3}{2}$	1	1	$\frac{13}{2}$

Therefore, C attains a minimum value of 13/2 when $x = 3/2$, $y = 1$, $u = \frac{1}{4}$ and $v = \frac{5}{4}$.

7. We first write the tableau

x	y	z	Const.
3	2	1	4
1	1	3	6
24	18	24	

Then obtain the following by interchanging rows and columns:

u	v	Const.
3	1	24
2	1	18
1	3	24
4	6	

From this table we construct the dual problem:

Maximize the objective function $P = 4u + 6v$ subject to

$$3u + v \leq 24$$
$$2u + v \leq 18$$
$$u + 3v \leq 24$$
$$u \geq 0, \ v \geq 0$$

The initial tableau is

u	v	x	y	z	P	Const.
3	1	1	0	0	0	24
2	1	0	1	0	0	18
1	3	0	0	1	0	24
−4	−6	0	0	0	1	0

Using the following sequence of row operations

1. $\frac{1}{3}R_3$ 2. $R_1 - R_3, \ R_2 - R_3, \ R_4 + 6R_3$ 3. $\frac{3}{8}R_1$ 4. $R_2 - \frac{5}{3}R_1, \ R_3 - \frac{1}{3}R_1, \ R_4 + 2R_1$

we obtain the final tableau

u	v	x	y	z	P	Const.
1	0	$\frac{3}{8}$	0	$-\frac{1}{8}$	0	6
0	0	$-\frac{5}{8}$	1	$-\frac{1}{8}$	0	0
0	1	$-\frac{1}{8}$	0	$\frac{3}{8}$	0	6
0	0	$\frac{3}{4}$	0	$\frac{7}{4}$	0	60

We conclude that C attains a minimum value of 60 when $x = 3/4, \ y = 0, \ z = 7/4, \ u = 6$, and $v = 6$.

9. Rewriting the problem, we have

Maximize $P = 3x - 4y$ subject to

$$x + y \leq 45$$
$$-x + 2y \leq -10$$
$$x \geq 0, \ y \geq 0$$

Using the simplex method, we have

x	y	u	v	P	Const		Ratio
1	1	1	0	0	45		45
-1	2	0	1	0	-10		10
-3	4	0	0	1	0		

$\xrightarrow{-R_2}$

x	y	u	v	P	Const
1	1	1	0	0	45
1	-2	0	-1	0	10
-3	4	0	0	1	0

$\xrightarrow{\substack{R_1-R_2 \\ R_3+3R_2}}$

x	y	u	v	P	Const
0	3	1	1	0	35
1	-2	0	-1	0	10
0	-2	0	-3	1	30

$\xrightarrow{\substack{R_2+R_1 \\ R_3+3R_1}}$

x	y	u	v	P	Const
0	3	1	1	0	35
1	1	1	0	0	45
0	7	3	0	1	135

We conclude that P attains a maximum value of 135 when $x = 45$ and $y = 0$.

11. We first write the problem in the form
Maximize $P = 2x + 3y$ subject to
$$2x + 5y \le 20$$
$$x - 5y \le -5$$
$$x \ge 0, \ y \ge 0$$
The initial tableau is

x	y	u	v	P	Const
2	5	1	0	0	20
1	-5	0	1	0	-5
-2	-3	0	0	1	0

Using the sequence of row operations
1. $\frac{1}{5}R_1$ 2. $R_2 + 5R_1$, $R_3 + 3R_1$ 3. $\frac{1}{3}R_2$ 4. $R_1 - \frac{2}{5}R_2$, $R_3 + \frac{4}{5}R_2$
we obtain the final tableau

x	y	u	v	P	Const
0	1	$\frac{1}{15}$	$-\frac{2}{15}$	0	2
1	0	$\frac{1}{3}$	$\frac{1}{3}$	0	5
0	0	$\frac{13}{15}$	$\frac{4}{15}$	1	16

We conclude that P attains a maximum value of 16 when $x = 5$ and $y = 2$.

13. Refer to Exercise 9, page 203. The problem simplifies to
Minimize $C = 14000x + 16000y$ subject to

$$2x + 3y \geq 26$$
$$3x + \ y \geq 18$$
$$x \geq 0, y \geq 0$$

We first write down the following tableaus for the primal problem:

x	y	$Const.$
2	3	26
3	1	18
14000	16000	

Next, we interchange the columns and rows of the foregoing tableau:

u	v	$Const.$
2	3	14000
3	1	16000
26	18	

This leads to the dual problem:

Maximize $P = 26u + 18v$ subject to
$$2u + 3v \leq 14000$$
$$3u + \ v \leq 16000$$
$$u \geq 0, v \geq 0$$

Let x and y be slack variables. We obtain the following tableaus:

	u	v	x	y	P	Const
	2	3	1	0	0	14000
p.r. →	$\boxed{3}$	1	0	1	0	16000
	−26	−18	0	0	1	0

p.c. (↑ under u)

Ratio
7000
$5333\frac{1}{3}$

$\frac{1}{3}R_2 \longrightarrow$

	u	v	x	y	P	Const
	2	3	1	0	0	14000
	1	$\frac{1}{3}$	0	$\frac{1}{3}$	0	$\frac{16000}{3}$
	−26	−18	0	0	1	0

$\begin{array}{l} R_1 - R_2 \\ R_3 + 26R_2 \end{array} \longrightarrow$

	u	v	x	y	P	Const
p.r. →	0	$\boxed{\frac{7}{3}}$	1	$-\frac{2}{3}$	0	$\frac{10000}{3}$
	1	$\frac{1}{3}$	0	$\frac{1}{3}$	0	$\frac{16000}{3}$
	0	$-\frac{28}{3}$	0	$\frac{26}{3}$	1	$\frac{41600}{3}$

p.c. (↑ under v)

Ratio
$\frac{10000}{7}$
16000

$\frac{3}{7}R_1 \longrightarrow$

	u	v	x	y	P	Const
	0	$\boxed{1}$	$\frac{3}{7}$	$-\frac{2}{7}$	0	$\frac{10000}{7}$
	1	$\frac{1}{3}$	0	$\frac{1}{3}$	0	$\frac{16000}{3}$
	0	$-\frac{28}{3}$	0	$\frac{26}{3}$	1	$\frac{416000}{3}$

	u	v	x	y	P	Const
$\xrightarrow{\begin{array}{c}R_3-\frac{1}{3}R_1\\ R_3-\frac{28}{3}R_1\end{array}}$	0	1	$\frac{3}{7}$	$-\frac{2}{7}$	0	$\frac{10000}{7}$
	1	0	$-\frac{1}{7}$	$\frac{3}{7}$	0	$\frac{34000}{7}$
	0	0	4	6	1	152000

The last tableau is in final form. The fundamental theorem of duality tells us that the solution to the primal problem is $x = 4$, $y = 6$, and $C = 152{,}000$. So, the Saddle Mine should be operated for 4 days and the Horseshoe Mine should be operated for 6 days at a minimum cost of $152,000/day.

15. Let x, y, and z denote the number of units of products A, B, and C made, respectively. Then the problem is to maximize the profit

$$P = 4x + 6y + 8z \text{ subject to}$$
$$9x + 12y + 18z \le 360$$
$$6x + 6y + 10z \le 240$$
$$x \ge 0, \ y \ge 0, \ z \ge 0$$

The initial tableau is

x	y	z	u	v	P	Const.
9	12	18	1	0	0	360
6	6	10	0	1	0	240
-4	-6	-8	0	0	1	0

Using the sequence of row operations

1. $\frac{1}{18}R_1$ 2. $R_2 - 10R_1$ 3. $R_3 + 8R_1$ 4. $\frac{3}{2}R_1$ 5. $R_2 + \frac{2}{3}R_1$, $R_3 + \frac{2}{3}R_1$

we obtain the final tableau

x	y	z	u	v	P	Const.
$\frac{3}{4}$	1	$\frac{3}{2}$	$\frac{1}{12}$	0	0	30
$\frac{3}{2}$	0	1	$-\frac{1}{2}$	1	0	60
$\frac{1}{2}$	0	1	$\frac{1}{2}$	0	1	180

and conclude that the company should produce 0 units of product A, 30 units of product B, and 0 units of product C to realize a maximum profit of $180. $13,000 on his investments.

CHAPTER 5

EXERCISES 5.1, page 290

1. The interest is given by $I = (500)(2)(0.08) = 80$, or $80.
    The accumulated amount is $500 + 80$, or $580.

3. The interest is given by $I = (800)(0.06)(0.75) = 36$, or $36.
    The accumulated amount is $800 + 36$, or $836.

5. We are given that $A = 1160$, $t = 2$, and $r = 0.08$, and we are asked to find P. Since
    $$A = P(1 + rt)$$
    we see that $\quad P = \dfrac{A}{1+rt} = \dfrac{1160}{1+(0.08)(2)} = 1000$, or $1000.

7. We use the formula $I = Prt$ and solve for t when $I = 20$, $P = 1000$, and $r = 0.05$.
    Thus,
    $$20 = 1000(0.05)(\tfrac{t}{365}), \quad \text{and} \quad t = \frac{365(20)}{50} = 146, \quad \text{or 146 days.}$$

9. We use the formula $A = P(1 + rt)$ with $A = 1075$, $P = 1000$, $t = 0.75$, and solve for r.
    Thus,
    $$1075 = 1000(1 + 0.75r)$$
    $$75 = 750r$$
    or $\qquad\qquad\qquad r = 0.10.$
    Therefore, the interest rate is 10 percent per year.

11. $A = 1000(1 + 0.07)^8 \approx 1718.19$, or $1718.19.

13. $A = 2500\left(1+\dfrac{0.07}{2}\right)^{20} \approx 4974.47$, or $4974.47.

15. $A = 12000\left(1+\dfrac{0.08}{4}\right)^{42} \approx 27{,}566.93$, or $27,566.93.

17. $A = 150,000\left(1+\dfrac{0.14}{12}\right)^{48} \approx 261,751.04$, or $261,751.04.

19. $A = 150,000\left(1+\dfrac{0.12}{365}\right)^{1095} = 214,986.69$, or $214,986.69.

21. Using the formula
$$r_{eff} = \left(1+\frac{r}{m}\right)^{m} - 1$$
with $r = 0.10$ and $m = 2$, we have
$$r_{eff} = \left(1+\frac{0.10}{2}\right)^{2} - 1 = 0.1025, \quad \text{or } 10.25 \text{ percent}..$$

23. Using the formula $\quad r_{eff} = \left(1+\dfrac{r}{m}\right)^{m} - 1$

with $r = 0.08$ and $m = 12$, we have
$$r_{eff} = \left(1+\frac{0.08}{12}\right)^{12} - 1 \approx 0.08300, \quad \text{or } 8.3 \text{ percent per year}.$$

25. The present value is given by $P = 40,000\left(1+\dfrac{0.08}{2}\right)^{-8} \approx 29,227.61$, or $29,227.61.

27. The present value is given by
$$P = 40,000\left(1+\frac{0.07}{12}\right)^{-48} \approx 30,255.95, \quad \text{or } \$30,255.95.$$

29. Think of $300 as the principal and $306 as the accumulated amount at the end of 30 days. If r denotes the simple interest rate per annum, then we have $P = 300$, $A = 306$, $t = 1/12$, and we are required to find r. Using (1b) we have
$$306 = 300\left(1+\frac{r}{12}\right) = 300 + r\left(\frac{300}{12}\right)$$
and $\quad r = \left(\dfrac{12}{300}\right)6 = 0.24$, or 24 percent per year.

31. The Abdullahs will owe $A = P(1 + rt) = 120,000[1 + (0.12)(\frac{3}{12})] = 123,600$, or $123,600.

33. Here $P = 10,000$, $I = 3500$, and $t = 7$, and so from Formula (1a), we have
$$3500 = 10000\,(r)\,7 \quad \text{and so } r = \frac{3500}{70000} = 0.05$$
So the bond pays interest at the rate of 5% per year.

35. The rate that you would expect to pay is
$$A = 380(1 + 0.08)^5 \approx 558.34, \text{ or } \$558.34 \text{ per day.}$$

37. The amount that they can expect to pay is given by
$$A = 210,000(1 + 0.05)^4 \approx 255,256, \text{ or approximately } \$255,256.$$

39. The investment will be worth $A = 1.5\left(1 + \dfrac{0.055}{2}\right)^{20} = 2.58064$, or approximately $2.58 million dollars.

41. We use Formula (3) with $P = 15,000$, $r = 0.098$, $m = 12$, and $t = 4$ giving the worth of Jodie's account as
$$A = 15,000\left(1 + \frac{0.098}{12}\right)^{(12)(4)} \approx 22,163.753, \text{ or approximately } \$22,163.75.$$

43. Using the formula $P = A\left(1 + \dfrac{r}{m}\right)^{-mt}$, we have
$$P = 40,000\left(1 + \frac{0.085}{4}\right)^{-20} \approx 26,267.49, \text{ or } \$26,267.49.$$

45. a. They should set aside
$$P = 100,000(1 + 0.085)^{-13} \approx 34,626.88, \text{ or } \$34,626.88.$$
   b. They should set aside
$$P = 100,000\left(1 + \frac{0.085}{2}\right)^{-26} \approx 33,886.16, \text{ or } \$33,886.16.$$

c. They should set aside

$$P = 100,000\left(1+\frac{0.085}{4}\right)^{-52} \approx 33,506.76, \text{ or } \$33,506.76.$$

47. The effective rate of interest for the Bendix Mutual Fund is

$$r_{eff} = \left(1+\frac{0.104}{4}\right)^4 - 1 \approx 0.1081 \quad \text{or} \quad 10.81\%/\text{yr}$$

whereas the effective rate of interest for the Acme Mutual fund is

$$r_{eff} = \left(1+\frac{0.106}{2}\right)^4 - 1 \approx 0.1088 \quad \text{or} \quad 10.88\%/\text{yr}$$

We conclude that the Acme Mutual Fund has a better rate of return.

49. The present value of the $8000 loan due in 3 years is given by

$$P = 8000\left(1+\frac{0.10}{2}\right)^{-6} - 1 = 5969.72, \text{ or } \$5969.72.$$

The present value of the $15,000 loan due in 6 years is given by

$$P = 15,000\left(1+\frac{0.10}{2}\right)^{-12} - 1 = 8352.56, \text{ or } \$8352.56.$$

Therefore, the amount the proprietors of the inn will be required to pay at the end of 5 years is given by

$$A = 14,322.28\left(1+\frac{0.10}{2}\right)^{10} = 23,329.48, \quad \text{or } \$23,329.48.$$

51. Using the compound interest formula with $A = 256,000$, $P = 200,000$ and $t = 6$, we have

$$256,000 = 200,000(1+R)^6$$
$$(1+R)^{1/6} = (1.28)^{1/6}$$
$$1+R = 1.042,$$
$$R = 0.042, \quad \text{or } 4.2 \text{ percent.}$$

53. Let the effective rate of interest be R. Then R satisfies
$$A = P(1+R)^t$$

or
$$10,000 = 6724.53\left(1+\frac{R}{2}\right)^{14}$$

$$1+\frac{R}{2} = (1.48709278)^{1/14} = 1.028750032$$

and $\quad R = 2(2.875) = 0.0575,$ or 5.75 percent.

55. Let the effective rate of interest be R. Then R satisfies
$$A = P(1 + R)^{t}$$
or $\quad 5170.42 = 5000(1 + R)^{245/365}$
$$1 + R = (1.034084)^{365/245} = 1.051199629$$
and $\quad R = 0.5119962,$ or approximately 5.12 percent.

57. Be definition, $A = P(1+r_{eff})^{t}$. So
$$(1+r_{eff})^{t} = \frac{A}{P}, \quad 1+r_{eff} = \left(\frac{A}{P}\right)^{1/t} \quad \text{and} \quad r_{eff} = \left(\frac{A}{P}\right)^{1/t} - 1.$$

59. False. Under compound interest $A = P(1+r)^{t}$ $(m = 1)$, whereas under simple interest, $A = P(1 + rt)$.

61. False. If Susan had gotten annual increases of 5 percent over 5 years, her salary would have been $A = 40,000(1+0.05)^{5} = 51,051.26$, or approximately $51,051 after 5 years and not $50,000.

USING TECHNOLOGY EXERCISES 5.1, page 295

1. $5872.78   3. $475.49   5. 8.95%/yr   7. 10.20%/yr

9. $29,743.30   11. $53,303.25

EXERCISES 5.2, page 304

1. $\quad S = 1000\left[\dfrac{(1+0.1)^{10}-1}{0.1}\right] = 15,937.42,$ or $15,937.42.

3. $S = 1800 \left[\dfrac{\left(1 + \dfrac{0.08}{4}\right)^{24} - 1}{\dfrac{0.08}{4}} \right] \approx 54{,}759.35$, or \$54,759.35.

5. $S = 600 \left[\dfrac{\left(1 + \dfrac{0.12}{4}\right)^{36} - 1}{\dfrac{0.12}{4}} \right] \approx 37{,}965.57$, or \$37,965.57.

7. $S = 200 \left[\dfrac{\left(1 + \dfrac{0.09}{12}\right)^{243} - 1}{\dfrac{0.09}{12}} \right] \approx 137{,}209.97$, or \$137,209.97.

9. $P = 5000 \left[\dfrac{1 - (1 + 0.08)^{-8}}{0.08} \right] \approx 28{,}733.19$, or \$28,733.19.

11. $P = 4000 \left[\dfrac{1 - (1 + 0.09)^{-5}}{0.09} \right] \approx 15{,}558.61$, or \$15,558.61.

13. $P = 800 \left[\dfrac{1 - \left(1 + \dfrac{0.12}{4}\right)^{-28}}{\dfrac{0.12}{4}} \right] \approx 15{,}011.29$, or \$15,011.29.

15. She will have
$$S = 1500 \left[\dfrac{(1 + 0.08)^{25} - 1}{0.08} \right] = 109{,}658.91, \text{ or } \$109{,}658.91.$$

17. On October 31, Linda's account will be worth

$$S = 40 \left[\frac{\left(1 + \frac{0.07}{12}\right)^{11} - 1}{\frac{0.07}{12}} \right] \approx 453.06, \text{ or } \$453.06.$$

One month later, this account will be worth $A = (453.06)\left(1 + \frac{0.07}{12}\right) = 455.70$, or $455.70.

19. The amount in Collin's employee retirement account is given by

$$S = 100 \left[\frac{\left(1 + \frac{0.07}{12}\right)^{144} - 1}{\frac{0.07}{12}} \right] \approx 22,469.50, \text{ or } \$22,469.50.$$

The amount in Collin's IRA is given by

$$S = 2000 \left[\frac{(1 + 0.09)^8 - 1}{0.09} \right] \approx 22,056.95, \text{ or } \$22,056.95.$$

Therefore, the total amount in his retirement fund is given by
22,469.50 + 22,056.95 = 44,526.45, or $44,526.45.

21. To find how much Karen has at age 65, we use formula (6) with $R = 150$,
$i = \frac{r}{m} = \frac{0.05}{12}$, and $n = mt = (12)(40) = 480$, giving

$$S = 150 \left[\frac{\left(1 + \frac{0.05}{12}\right)^{480} - 1}{\frac{0.05}{12}} \right] \approx 228,903.0235$$

or $228,903.02. To find how much Matt will have upon attaining the age of 65, use
Formula (6) with $R = 250$, $i = \frac{r}{m} = \frac{0.05}{12}$, and $n = mt = (12)(30) = 360$ giving

$$S = 250 \left[\frac{\left(1 + \dfrac{0.05}{12}\right)^{360} - 1}{\dfrac{0.05}{12}} \right] \approx 208,064.6588, \text{ or } \$208,064.66. \text{ So Karen}$$

will have the bigger nest egg.

23. The equivalent cash payment is given by

$$P = 450 \left[\frac{1 - \left(1 + \dfrac{0.09}{12}\right)^{-24}}{\dfrac{0.09}{12}} \right] \approx 9850.12, \text{ or } \$9850.12.$$

25. We use the formula for the present value of an annuity obtaining

$$P = 22 \left[\frac{1 - \left(1 + \dfrac{0.18}{12}\right)^{-36}}{\dfrac{0.18}{12}} \right] \approx 608.54, \text{ or } \$608.54.$$

27. With an $1200 monthly payment, the present value of their loan would be

$$P = 1200 \left[\frac{1 - \left(1 + \dfrac{0.095}{12}\right)^{-360}}{\dfrac{0.095}{12}} \right] \approx 142,712.02, \text{ or } \$142,712.02.$$

With a $1500 monthly payment, the present value of their loan would be

$$P = 1500 \left[\frac{1 - \left(1 + \dfrac{0.095}{12}\right)^{-360}}{\dfrac{0.095}{12}} \right] \approx 178,390.02 \text{ or } \$178,390.02.$$

Since they intend to make a $25,000 down payment, the range of homes they should consider is $167,712 to $203,390.

29. The lower limit of their investment is

$$A = 1200 \left[\dfrac{1 - \left(1 + \dfrac{0.09}{12}\right)^{-180}}{\dfrac{0.09}{12}} \right] + 25{,}000 \approx 143{,}312.09$$

or approximately \$143,312. The upper limit of their investment is

$$A = 1500 \left[\dfrac{1 - \left(1 + \dfrac{0.09}{12}\right)^{-180}}{\dfrac{0.09}{12}} \right] + 25{,}000 \approx 172{,}890.11$$

or approximately \$172,890. Therefore, the price range of houses they should consider is \$143,312 to \$172,890.

31. False. This statement would be true only if the interest rate is equal to zero.

USING TECHNOLOGY EXERCISES 5.2, page 308

1. \$59,622.15   3. \$8453.59   5. \$35,607.23   7. \$13,828.60

EXERCISES 5.3, page 315

1. The size of each installment is given by

$$R = \dfrac{100{,}000(0.08)}{1 - (1 + 0.08)^{-10}} \approx 14{,}902.95, \text{ or } \$14{,}902.95.$$

3. The size of each installment is given by

$$R = \dfrac{5000(0.01)}{1 - (1 + 0.01)^{-12}} \approx 444.24, \text{ or } \$444.24.$$

5. The size of each installment is given by

$$R = \dfrac{25{,}000(0.0075)}{1 - (1 + 0.0075)^{-48}} \approx 622.13, \text{ or } \$622.13.$$

7. The size of each installment is
$$R = \frac{80{,}000(0.00875)}{1-(1+0.00875)^{-360}} \approx 731.79, \text{ or } \$731.79.$$

9. The periodic payment that is required is
$$R = \frac{20{,}000(0.02)}{(1+0.02)^{12}-1} \approx 1491.19, \text{ or } \$1491.19.$$

11. The periodic payment that is required is
$$R = \frac{100{,}000(0.0075)}{(1+0.0075)^{120}-1} \approx 516.76, \text{ or } \$516.76$$

13. The periodic payment that is required is
$$R = \frac{250{,}000(0.00875)}{(1+0.00875)^{300}-1} \approx 172.95, \text{ or } \$172.95.$$

15. The periodic payment that is required is
$$R = \frac{50000\left(\dfrac{.10}{4}\right)}{\left(1+\dfrac{.10}{4}\right)^{20}-1} = 1957.36, \text{ or } \$1957.36.$$

17. The periodic payment that is required is
$$R = \frac{35000\left(\dfrac{0.075}{2}\right)}{1-\left(1+\dfrac{0.075}{2}\right)^{-13}} = 3450.87, \text{ or } \$3450.87.$$

19. The size of each installment is given by
$$R = \frac{100{,}000(0.10)}{1-(1+0.10)^{-10}} \approx 16{,}274.54, \text{ or } \$16{,}274.54.$$

21. The monthly payment in each case is given by

$$R = \frac{100{,}000\left(\dfrac{r}{12}\right)}{1-\left(1+\dfrac{r}{12}\right)^{-360}}$$

Thus, if $r = 0.08$, then $R = \dfrac{100{,}000\left(\dfrac{0.08}{12}\right)}{1-\left(1+\dfrac{0.08}{12}\right)^{-360}} \approx 733.76$, or $733.76

If $r = 0.09$, then $R = \dfrac{100{,}000\left(\dfrac{0.09}{12}\right)}{1-\left(1+\dfrac{0.09}{12}\right)^{-360}} \approx 804.62$, or $804.62

If $r = 0.10$, then $R = \dfrac{100{,}000\left(\dfrac{0.10}{12}\right)}{1-\left(1+\dfrac{0.10}{12}\right)^{-360}} \approx 877.57$, or $877.57.

If $r = 0.11$, then $R = \dfrac{100{,}000\left(\dfrac{0.11}{12}\right)}{1-\left(1+\dfrac{0.11}{12}\right)^{-360}} \approx$ $952.32.

a. The difference in monthly payments in the two loans is
$877.57 - $665.30 = $212.27.
b. The monthly mortgage payment on a $150,000 mortgage would be
1.5($877.57) = $1316.36.
The monthly mortgage payment on a $50,000 mortgage would be
0.5($877.57) = $438.79.

23. a. The amount of the loan required is 16000 - (0.25)(16000) or 12,000 dollars. If the car is financed over 36 months, the payment will be

$$R = \frac{12,000\left(\dfrac{0.10}{12}\right)}{1-\left(1+\dfrac{0.10}{12}\right)^{-36}} \approx 387.21, \text{ or } \$387.21 \text{ per month.}$$

If the car is financed over 48 months, the payment will be

$$R = \frac{12,000\left(\dfrac{0.10}{12}\right)}{1-\left(1+\dfrac{0.10}{12}\right)^{-48}} \approx 304.35, \text{ or } \$304.35 \text{ per month.}$$

b. The interest charges for the 36-month plan are
$$36(387.21) - 12000 = 1939.56,$$
or $1939.56. The interest charges for the 48-month plan are
$$48(304.35) - 12000 = 2608.80, \text{ or } \$2608.80.$$

25. The amount borrowed is 270,000 - 30,000 = 240,000 dollars. The size of the monthly installment is

$$R = \frac{240,000\left(\dfrac{0.08}{12}\right)}{1-\left(1+\dfrac{0.08}{12}\right)^{-360}} \approx 1761.03, \text{ or } \$1761.03.$$

To find their equity after five years, we compute

$$P = 1761.03\left[\frac{1-\left(1+\dfrac{0.08}{12}\right)^{-300}}{\dfrac{0.08}{12}}\right] \approx 228,167$$

or $228,167, and so their equity is 270,000 − 228,167 = 41,833, or $41,833. To find their equity after ten years, we compute

$$P = 1761.03\left[\frac{1-\left(1+\dfrac{0.08}{12}\right)^{-240}}{\dfrac{0.08}{12}}\right] \approx 210,539, \text{ or } \$210,539.$$

and their equity is $270,000 - 210,539 = 59,461$, or \$59,461.

To find their equity after twenty years, we compute

$$P = 1761.03 \left[\frac{1 - \left(1 + \dfrac{0.08}{12}\right)^{-120}}{\dfrac{0.08}{12}} \right] \approx 145,147, \text{ or } \$145,147,$$

and their equity is $270,000 - 145,147$, or \$124,853.

27. The amount that must be deposited quarterly into this fund is

$$R = \frac{\left(\dfrac{0.09}{4}\right)200,000}{\left(1 + \dfrac{0.09}{4}\right)^{40} - 1} \approx 3,135.48, \text{ or } \$3,135.48.$$

29. The size of each quarterly installment is given by

$$R = \frac{\left(\dfrac{0.10}{4}\right)20,000}{\left(1 + \dfrac{0.10}{4}\right)^{12} - 1} \approx 1449.74, \text{ or } \$1449.74.$$

31. The size of each monthly installment is given by

$$R = \frac{\left(\dfrac{0.085}{12}\right)250,000}{\left(1 + \dfrac{0.085}{12}\right)^{300} - 1} \approx 242.23, \text{ or } \$242.23.$$

33. The value of the IRA account after 20 years is

$$S = 375 \left[\frac{\left(1 + \dfrac{0.08}{4}\right)^{80} - 1}{\dfrac{0.08}{4}} \right] \approx 72,664.48, \text{ or } \$72,664.48.$$

The payment he would receive at the end of each quarter for the next 15 years is given by $R = \dfrac{\left(\dfrac{0.08}{4}\right)72{,}664.48}{1-\left(1+\dfrac{0.08}{4}\right)^{-60}} \approx 2090.41$, or \$2090.41.

If he continues working and makes quarterly payments until age 65, the value of the IRA account would be

$$S = 375\left[\dfrac{\left(1+\dfrac{0.08}{4}\right)^{100}-1}{\dfrac{0.08}{4}}\right] \approx 117{,}087.11, \text{ or } \$117{,}087.11.$$

The payment he would receive at the end of each quarter for the next 10 years is

given by $R = \dfrac{\left(\dfrac{0.08}{4}\right)117{,}087.11}{1-\left(1+\dfrac{0.08}{4}\right)^{-40}} \approx 4280.21$, or \$4280.21.

35. The monthly payment the Sandersons are required to make under the terms of their original loan is given by

$$R = \dfrac{100{,}000\left(\dfrac{0.10}{12}\right)}{1-\left(1+\dfrac{0.10}{12}\right)^{-240}} \approx 965.02, \text{ or } \$965.02.$$

The monthly payment the Sandersons are required to make under the terms of their new loan is given by

$$R = \dfrac{100{,}000\left(\dfrac{0.078}{12}\right)}{1-\left(1+\dfrac{0.078}{12}\right)^{-240}} \approx 824.04, \text{ or } \$824.04.$$

The amount of money that the Sandersons can expect to save over the life of the loan by refinancing is given by $240(965.02 - 824.04) = 33{,}835.20$, or \$33,835.20.

37. Sarah's monthly payment is found using Formula (10) with $P = 200{,}000$,

$i = \dfrac{r}{m} = \dfrac{0.06}{12}$, and $n = (12)(15) = 180$. Thus,

$$R = 200{,}000 \left[\dfrac{\dfrac{0.06}{12}}{1 - \left(1 + \dfrac{0.06}{12}\right)^{-360}} \right] \approx 1687.7137. \quad \text{After 85years, she has paid}$$

(5)(12) or 60 payments. Her outstanding principal is given by the sum of the remaining installments, (180 – 60), or 120. Using Formula 8, we find

$$P = 1687.7137 \left[\dfrac{1 - \left(1 + \dfrac{0.06}{12}\right)^{-120}}{\dfrac{0.06}{12}} \right] \approx 152{,}018.2012$$

So her outstanding principal is $152,018.20.

39. a. Here $P = 200{,}000$, $i = \dfrac{r}{m} = \dfrac{0.095}{12}$, and $n = mt = (12)(30) = 360$. Therefore,

$$R = \dfrac{200{,}000 \left(\dfrac{0.095}{12}\right)}{1 - \left(1 + \dfrac{0.095}{12}\right)^{-360}} \approx 1681.7084, \text{ and so her monthly payment is } \$1681.71.$$

b. After (4)(12) = 48 monthly payments have been made, her outstanding principal is given by the sum of the present values of the remaining installments (which is 360 – 48 = 312). Using Formula (8), we find it to be

$$P = 1681.7084 \left[\dfrac{1 - \left(1 + \dfrac{0.095}{12}\right)^{-312}}{\dfrac{0.095}{12}} \right] \approx 194{,}282.6675,$$

or approximately $194,282.67.

c. Here $P = 194{,}282.6675$, $i = \dfrac{r}{m} = \dfrac{0.0675}{12}$, and $n = (12)(30) = 360$. So

$$R = \dfrac{194282.67 \left(\dfrac{0.0675}{12}\right)}{1 - \left(1 + \dfrac{0.0675}{12}\right)^{-360}} \approx 1260.1137$$

and so her new monthly payment is $1260.11.

d. Emily will save 1681.71 − 1260.11, or $421.60 per month.

41. The amount of the loan the Meyers need to secure is $280,000. Using the bank's financing, the monthly payment would be

$$R = \frac{280,000\left(\dfrac{0.11}{12}\right)}{1 - \left(1 + \dfrac{0.11}{12}\right)^{-300}} \approx 2744.32, \text{ or } \$2744.32.$$

Using the seller's financing, the monthly payment would be

$$R = \frac{280,000\left(\dfrac{0.098}{12}\right)}{1 - \left(1 + \dfrac{0.098}{12}\right)^{-300}} \approx 2504.99, \text{ or } \$2504.99.$$

By choosing the seller's financing rather than the bank's, the Meyers would save $(2744.32 - 2504.99)(300) = 71,799$, or $71,799 in interest.

USING TECHNOLOGY EXERCISES 5.3, page 305

1. $3645.40   3. $18,443.75   5. $1863.99   7. $707.96

9. $18,288.92. The amortization schedule follows.

End of Period	Interest charged	Repayment made	Payment toward Principal	Outstanding Principal
0				120,000.00
1	10,200.00	18,288.92	8,088.92	111,911.08
2	9,512.44	18,288.92	8,776.48	103,134.60
3	8,766.44	18,288.92	9,522.48	93,612.12
4	7,957.03	18,288.92	10,331.89	83,280.23
5	7,078.82	18,288.92	11,210.10	72,070.13
6	6,125.96	18,288.92	12,162.96	59,907.17
7	5,092.11	18,288.92	13,196.81	46,710.36
8	3,970.38	18,288.92	14,318.54	32,391.82
9	2,753.30	18,288.92	15,535.62	16,856.20
10	1,432.78	18,288.98	16,856.14	.00

EXERCISES 5.4, page 312

1. $a_9 = 6 + (9 - 1)3 = 30$

3. $a_8 = -15 + (8 - 1)\left(\frac{3}{2}\right) = -\frac{9}{2} = -4.5.$

5. $a_{11} - a_4 = (a_1 + 10d) - (a_1 + 3d) = 7d.$ Also, $a_{11} - a_4 = 107 - 30 = 77.$
    Therefore, $7d = 77,$ and $d = 11.$ Next,
    $$a_4 = a + 3d = a + 3(11) = a + 33 = 30.$$
    and $a = -3.$ Therefore, the first five terms are -3, 8, 19, 30, 41.

7. Here $a = x,$ $n = 7,$ and $d = y.$ Therefore, the required term is
    $$a_7 = x + (7 - 1)y = x + 6y.$$

9. Using the formula for the sum of the terms of an arithmetic progression with $a = 4,$
    $d = 7$ and $n = 15,$ we have
    $$S_n = \frac{n}{2}[2a + (n-1)d]]$$
    $$S_{15} = \frac{15}{2}[2(4) + (15-1)7] = \frac{15}{2}(106) = 795$$

11. The common difference is $d = 2$ and the first term is $a = 15.$ Using the formula for
    the nth term
    $$a_n = a + (n - 1)d,$$
    we have $\quad 57 = 15 + (n - 1)(2) = 13 + 2n$
    $\quad 2n = 44,$ and $\quad n = 22.$
    Using the formula for the sum of the terms of an arithmetic progression with $a = 15,$
    $d = 2$ and $n = 22,$ we have
    $$S_n = \frac{n}{2}[2a + (n-1)d]]$$
    $$S_{22} = \frac{22}{2}[2(15) + (22-1)2] = 11(72) = 792.$$

13. $\quad f(1) + f(2) + f(3) + \cdots + f(20)$
    $$= [3(1) - 4] + [3(2) - 4] + [3(3) - 4] + \cdots + [3(20) - 4]$$
    $$= 3(1 + 2 + 3 + \cdots + 20) + 20(-4)$$

$$= 3\left(\frac{20}{2}\right)[2(1) + (20-1)1] - 80$$
$$= 550.$$

15. $$S_n = \frac{n}{2}[2a_1 + (n-1)d]] = \frac{n}{2}(a_1 + a_1 + (n-1)d]$$
$$= \frac{n}{2}(a_1 + a_n)$$

a. $S_{11} = \frac{11}{2}(3 + 47) = 275.$ b. $S_{20} = \frac{20}{2}[5 + (-33)] = -280.$

17. Let n be the number of weeks till she reaches 10 miles. Then
$$a_n = 1 + (n-1)\frac{1}{4} = 1 + \frac{1}{4}n - \frac{1}{4} = \frac{1}{4}n + \frac{3}{4} = 10$$
Therefore, $n + 3 = 40$, and $n = 37$; that is, it will take Karen 37 weeks to meet her goal.

19. To compute the Kunwoo's fare by taxi, take $a = 1$, $d = 0.60$, and $n = 25$. Then the required fare is given by
$$a_{25} = 1 + (25 - 1)0.60 = 15.4,$$
or $15.40. Therefore, by taking the airport limousine, the Kunwoo will save
15.40 - 7.50 = 7.90, or $7.90.

21. a. Using the formula for the sum of an arithmetic progression, we have
$$S_n = \frac{n}{2}[2a + (n-1)d]]$$
$$= \frac{N}{2}[2(1) + (N-1)(1)] = \frac{N}{2}(N+1).$$
   b. $S_{10} = \frac{10}{2}(10 + 1) = 5(11) = 55$
$$D_3 = (C-S)\frac{N - (n-1)}{S_N} = (6000 - 500)\frac{10 - (3-1)}{55} = 5500\left(\frac{8}{55}\right)$$
$$= 800, \text{ or } \$800.$$

23. This is a geometric progression with $a = 4$ and $r = 2$. Next, $a_7 = 4(2)^6 = 256$,
   and $$S_7 = \frac{4(1 - 2^7)}{1 - 2} = 508.$$

25. If we compute the ratios
$$\frac{a_2}{a_1} = \frac{-\frac{3}{8}}{\frac{1}{2}} = -\frac{3}{4} \quad \text{and} \quad \frac{a_3}{a_2} = \frac{\frac{1}{4}}{-\frac{3}{8}} = -\frac{2}{3},$$
we see that the given sequence is not geometric since the ratios are not equal.

27. This is a geometric progression with $a = 243$, and $r = 1/3$.
$$a_7 = 243(\tfrac{1}{3})^6 = \tfrac{1}{3}$$
$$S_7 = \frac{243(1-(\tfrac{1}{3})^7)}{1-\tfrac{1}{3}} = 364\tfrac{1}{3}.$$

29. First, we compute $\quad r = \dfrac{a_2}{a_1} = \dfrac{3}{-3} = -1.$

Next, $a_{20} = -3(-1)^{19} = 3$ and so $S_{20} = \dfrac{-3[1-(-1)^{20}]}{[1-(-1)]} = 0.$

31. The population in five years is expected to be
$$200,000(1.08)^{6-1} = 200,000\,(1.08)^5 \approx 293,866.$$

33. The salary of a union member whose salary was $32,000 six years ago is given by

the 7th term of a geometric progression whose first term is 32,000 and whose common ratio is 1.03. Thus
$$a_7 = (32,000)(1.03)^6 = 38,209.67, \text{ or } \$38,209.67.$$

35. With 8 percent raises per year, the employee would make
$$S_4 = 38,000\left[\frac{1-(1.08)^4}{1-1.08}\right] \approx 171,232.26,$$
or $171,232.26 over the next four years.
With $2500 raises per year, the employee would make
$$S_4 = \frac{4}{2}[2(38,000)+(4-1)2500] = 167,000$$
or $167,000 over the next four years. We conclude that the employee should choose annual raises of 8 percent per year.

37. a. During the sixth year, she will receive
$$a_6 = 10{,}000(1.15)^5 \approx 20{,}113.57, \text{ or } \$20{,}113.57.$$
b. The total amount of the six payments will be given by
$$S_6 = \frac{10{,}000[1-(1.15)^6]}{1-1.15} \approx 87{,}537.38, \text{ or } \$87{,}537.38.$$

39. The book value of the office equipment at the end of the eighth year is given by
$$V(8) = 150{,}000\left(1-\frac{2}{10}\right)^8 \approx 25{,}165.82, \text{ or } \$25{,}165.82.$$

41. The book value of the restaurant equipment at the end of six years is given by
$$V(6) = 150{,}000(0.8)^6 \approx 39{,}321.60,$$
or $39,321.60. By the end of the sixth year, the equipment will have depreciated by
$$D(n) = 150{,}000 - 39{,}321.60 = 110{,}678.40, \text{ or } \$110{,}678.40.$$

43. True. Suppose d is the common difference of $a_1, a_2, ..., a_n$ and e is the common difference of $b_1, b_2, ..., b_n$. Then $d + e$ is the common difference of $a_1 + b_1, a_2 + b_2, ..., a_n + b_n$ is obtained by adding the constant $c + d$ to it, we see that it is indeed an arithmetic progression.

CHAPTER 5, REVIEW EXERCISES, page 331

1. a. Here $P = 5000$, $r = 0.1$, and $m = 1$. Thus, $i = r = 0.1$ and $n = 4$. So
$$A = 5000(1.1)^4 = 7320.5, \text{ or } \$7320.50.$$
    b. Here $m = 2$ so that $i = 0.1/2 = 0.05$ and $n = (4)(2) = 8$. So
$$A = 5000(1.05)^8 \approx 7387.28 \text{ or } \$7387.28.$$
    c. Here $m = 4$, so that $i = 0.1/4 = 0.025$ and $n = (4)(4) = 16$. So
$$A = 5000(1.025)^{16} \approx 7{,}422.53, \text{ or } \$7422.53.$$
    d. Here $m = 12$, so that $i = 0.1/12$ and $n = (4)(12) = 48$. So
$$A = 5000\left(1+\frac{0.10}{12}\right)^{48} \approx 7446.77, \text{ or } \$7446.77.$$

3. a. The effective rate of interest is given by

$$r_{eff} = \left(1 + \frac{r}{m}\right)^m - 1 = (1 + 0.12) - 1 = 0.12, \text{ or } 12 \text{ percent.}$$

b. The effective rate of interest is given by

$$r_{eff} = \left(1 + \frac{r}{m}\right)^m - 1 = \left(1 + \frac{0.12}{2}\right)^2 - 1 = 0.1236, \text{ or } 12.36 \text{ percent.}$$

c. The effective rate of interest is given by

$$r_{eff} = \left(1 + \frac{r}{m}\right)^m - 1 = \left(1 + \frac{0.12}{4}\right)^4 - 1 \approx 0.125509, \text{ or } 12.5509 \text{ percent.}$$

d. The effective rate of interest is given by

$$r_{eff} = \left(1 + \frac{r}{m}\right)^m - 1 = \left(1 + \frac{0.12}{12}\right)^{12} - 1 \approx 0.126825, \text{ or } 12.6825 \text{ percent.}$$

5. The present value is given by
$$P = 41{,}413\left(1 + \frac{0.065}{4}\right)^{-20} \approx 30{,}000.29, \quad \text{or approximately } \$30{,}000.$$

7. $$S = 150\left[\frac{\left(1 + \frac{0.08}{4}\right)^{28} - 1}{\frac{0.08}{4}}\right] \approx 5557.68, \text{ or } \$5557.68.$$

9. Using the formula for the present value of an annuity with $R = 250$, $n = 36$,

$i = 0.09/12 = 0.0075$, we have
$$P = 250\left[\frac{1 - (1.0075)^{-36}}{0.0075}\right] \approx 7861.70, \text{ or } \$7861.70.$$

11. Using the amortization formula with $P = 22{,}000$, $n = 36$, and $i = 0.085/12$, we find

$$R = \frac{22,000\left(\dfrac{0.085}{12}\right)}{1-\left(1+\dfrac{0.085}{12}\right)^{-36}} \approx 694.49, \text{ or } \$694.49.$$

13. Using the sinking fund formula with $S = 18,000$, $n = 48$, and $i = 0.06/12$, we have

$$R = \frac{\left(\dfrac{0.06}{12}\right)18,000}{\left(1+\dfrac{0.06}{12}\right)^{48}-1} \approx 332.73, \text{ or } \$332.73.$$

15. We are asked to find r such that

$$\left(1+\frac{r}{365}\right)^{365} = \left(1+\frac{0.072}{12}\right)^{12} = 1.074424168.$$

Then
$$1+\frac{r}{365} = (1.074424168)^{1/365}$$

$$\frac{r}{365} = (1.074424168)^{1/365} - 1$$

and
$$r = 365[1.074424168^{1/365} - 1]$$
$$\approx 0.071791919, \text{ or approximately } 7.179 \text{ percent.}$$

17. Let a_n denote the sales during the nth year of operation. Then

$$\frac{a_{n+1}}{a_n} = 1.14 = r.$$

Therefore, the sales during the fourth year of operation were
$$a_4 = a_1 r^{n-1} = 1,750,000(1.14)^{4-1} = 2,592,702, \text{ or } \$2,592,702.$$
The total sales over the first four years of operation are given by

$$S_4 = \frac{a(1-r^4)}{1-r} = \frac{1,750,000[1-(1.14)^4]}{1-(1.14)} = 8,612,002, \text{ or } \$8,612,002.$$

19. Using the present value formula for compound interest, we have

$$P = A\left(1+\frac{r}{m}\right)^{-mt} = 19,440.31\left(1+\frac{0.065}{12}\right)^{-12(4)} = 15,000.00, \text{ or } \$15,000.$$

21. Using the sinking fund formula with $S = 40,000$, $n = 120$, and $i = 0.08/12$, we find

$$R = \frac{\left(\dfrac{0.08}{12}\right)40,000}{\left(1+\dfrac{0.08}{12}\right)^{120}-1} = 218.64, \text{ or } \$218.64.$$

23. Using the formula for the present value of an annuity, we see that the equivalent cash payment of Maria's auto lease is

$$P = 300\,\frac{1-\left(1+\dfrac{0.05}{12}\right)^{-48}}{\dfrac{0.05}{12}} = 13,026.89, \text{ or } \$13,026.89.$$

25. a. The monthly payment is given by
$$P = \frac{(120,000)(0.0075)}{1-(1+0.0075)^{-360}} \approx 965.55, \text{ or } \$965.55.$$

    b. We can find the total interest payment by computing
        $360(965.55) - 120,000 \approx 227,598, \text{ or } \$227,598.$
    c. We first compute the present value of their remaining payments. Thus,
$$P = 965.55\left[\frac{1-(1+0.0075)^{-240}}{0.0075}\right] \approx 107,316.01.$$

    or $107,316.01. Then their equity is
        $150,000 - 107,316.01, \text{ or approximately } \$42,684.$

27. Using the sinking fund formula with S = 500,000, n = 20, and I = 0.10/4, we find that the amount of each installment should be

$$R = \frac{\left(\dfrac{0.10}{4}\right)500,000}{\left(1+\dfrac{0.10}{4}\right)^{20}-1} \approx 19,573.56, \text{ or } \$19,573.56.$$

29. Using the amortization formula, we find that Bill's monthly payment will be

$$R = \frac{3200\left(\dfrac{0.186}{12}\right)}{1-\left(1+\dfrac{0.186}{12}\right)^{-18}} \approx 205.09, \text{ or } \$205.09.$$

CHAPTER 6

EXERCISES 6.1, page 342

1. $\{x \mid x$ is gold medalist in the 2002 Winter Olympic Games$\}$

3. $\{x \mid x$ is an integer greater than 2 and less than 8$\}$

5. $\{2,3,4,5,6\}$ 7. $\{-2\}$

9. a. True--the order in which the elements are listed is not important.
    b. False-- A is a set, not an element.

11. a. False. The empty set has no elements. b. False. 0 is an element and $\varnothing$ is a set.

13. True.

15. a. True. 2 belongs to A. b. False. For example, 5 belongs to A but 5 $\notin$ $\{2,4,6\}$.

17. a. and b.

19. a. $\varnothing$, $\{1\}$, $\{2\}$, $\{1,2\}$
    b. $\varnothing$, $\{1\}$, $\{2\}$, $\{3\}$, $\{1,2\}$, $\{1,3\}$, $\{2,3\}$, $\{1,2,3\}$
    c. $\varnothing$, $\{1\}$, $\{2\}$, $\{3\}$, $\{4\}$, $\{1,2\}$, $\{1,3\}$, $\{1,4\}$, $\{2,3\}$, $\{2,4\}$, $\{3,4\}$, $\{1,2,3\}$,
       $\{1,2,4\}$, $\{2,3,4\}$, $\{1,3,4\}$, $\{1,2,3,4\}$

21. $\{1, 2, 3, 4, 6, 8, 10\}$ 23. $\{$Jill, John, Jack, Susan, Sharon$\}$

25. a.   b.   c.

27. a.

b.

29. a.

b.

31. a.

b.

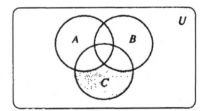

33. a. $A^C = \{2,4,6,8,10\}$
    b. $B \cup C = \{2,4,6,8,10\} \cup \{1,2,4,5,8,9\} = \{1,2,4,5,6,8,9,10\}$
    c. $C \cup C^C = U = \{1,2,3,4,5,6,7,8,9,10\}$

35. a. $(A \cap B) \cup C = C = \{1,2,4,5,8,9\}$
    b. $(A \cup B \cup C)^C = \varnothing$
    c. $(A \cap B \cap C)^C = U = \{1,2,3,4,5,6,7,8,9,10\}$

37. a. The sets are not disjoint. 4 is an element of both sets.
    b. The sets are disjoint as they have no common elements.

39. a. The set of all employees at the Universal Life Insurance Company who do not drink tea.
b. The set of all employees at the Universal Life Insurance Company who do not drink coffee.

41. a. The set of all employees at the Universal Life Insurance Company who drink tea but not coffee.
b. The set of all employees at the Universal Life Insurance Company who drink coffee but not tea.

43. a. The set of all employees at the hospital who are not doctors
b. The set of all employees at the hospital who are not nurses

45. a. The set of all employees at the hospital who are female doctors.
b. The set of all employees at the hospital who are both doctors and administrators.

47. a. $D \cap F$ b. $R \cap F^C \cap L^C$

49. a. B^C b. $A \cap B$ c. $A \cap B \cap C^C$

51. a. Region 1: $A \cap B \cap C$ is the set of tourists who used all three modes of transportation over a 1-week period in London.
b. Regions 1 and 4: $A \cap C$ is the set of tourists who have taken the underground and a bus over a 1-week period in London.
c. Regions 4, 5, 7, and 8: B^c is the set of tourists who have not taken a cab over a 1-week period in London.

53. $A \subset A \cup B$ $B \subset A \cup B$

    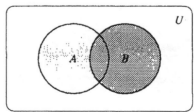

55. $A \cup (B \cup C) = (A \cup B) \cup C$

   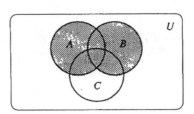

57. $A \cap (B \cup C) = (A \cap B) \cup (A \cap C)$

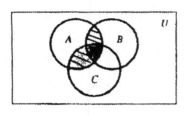

59. a. $A \cup (B \cup C) = \{1,3,5,7,9\} \cup (\{1,2,4,7,8\} \cup \{2,4,6,8\})$
$$= \{1\ 3,5,7,9\} \cup \{1,2,4,6,7,8\}$$
$$= \{1,2,3,4,5,6,7,8,9\}$$
$(A \cup B) \cup C = (\{1,3,5,7,9\} \cup (\{1,2,4,7,8\}) \cup \{2,4,6,8\})$
$$= \{1,2,3,4,5,7,8,9\} \cup \{2,4,6,8\}$$
$$= \{1,2,3,4,5,6,7,8,9\}$$
  b. $A \cap (B \cap C) = \{1,3,5,7,9\} \cap (\{1,2,4,7,8\} \cap \{2,4,6,8\})$
$$= \{1,3,5,7,9\} \cap (\{2,4,8\}$$
$$= \varnothing$$
$(A \cap B) \cap C = (\{1,3,5,7,9\} \cap \{1,2,4,7,8\}) \cap \{2,4,6,8\}$
$$= \{1,7\} \cap \{2,4,6,8\}$$
$$= \varnothing.$$

61. a. r, u, v, w, x, y b. v, r 63. a. t, y, s b. t, s, w, x, z 65. $A \subset C$

67. False. Since every element in a set A belongs to A, A is a subset of itself.

69. True. If at least one of the sets A or B is nonempty, then $A \cup B \neq \varnothing$.

71. True. $(A \cup A^c)^c = U^c = \varnothing$.

EXERCISES 6.2, page 350

1. $A \cup B = \{a,e,g,h,i,k,l,m,o,u\}$, and so $n(A \cup B) = 10$.   Next,
   $n(A) + n(B) = 5 + 5 = 10$.

3. a. $A = \{2,4,6,8\}$ and $n(A) = 4$.   b. $B = \{6,7,8,9,10\}$ and $n(B) = 5$
   c. $A \cup B = \{2,4,6,7,8,9,10\}$ and $n(A \cup B) = 7$.
   d. $A \cap B = \{6,8\}$ and $n(A \cap B) = 2$.

5. Using the results of Exercise 3, we see that $n(A \cup B) = 7$ and
   $n(A) + n(B) - n(A \cap B) = 4 + 5 - 2 = 7$.

7. Since $n(A \cup B) = n(A) + n(B) - n(A \cap B)$
   $n(B) = n(A \cup B) + n(A \cap B) - n(A)$
   $= 30 + 5 - 15 = 20$.

9. Refer to the Venn diagram at the right.
   a. $n(A \cup B) = 60 + 40 + 40 = 140$.
   b. $n(A^C) = 40 + 60 = 100$.
   c. $n(A \cap B^C) = 60$.

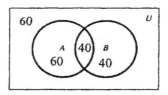

11. $n(A \cup B) = n(A) + n(B) - n(A \cap B) = 6 + 10 - 3 = 13$.

13. $n(A \cup B) = n(A) + n(B) - n(A \cap B)$ so
    $n(A \cap B) = n(A) + n(B) - n(A \cup B) = 4 + 5 - 9 = 0$.

15. $n(A \cap B \cap C)$
    $= n(A \cup B \cup C) - n(A) - n(B) - n(C) + n(A \cap B) + n(A \cap C) + n(B \cap C)$
    so $n(C) = n(A \cup B \cup C) - n(A \cap B \cap C) - n(A) - n(B)$
    $+ n(A \cap B) + n(A \cap C) + n(B \cap C$
    $= 25 - 2 - 12 - 12 + 5 + 5 + 4 = 13$.

17. Let A denote the set of prisoners in the Wilton County Jail who were accused of
    a felony and B the set of prisoners in that jail who were accused of a misdemeanor.
    Then we are given that
    $n(A \cup B) = 190$

Refer to the diagram at the right.
Then the number of prisoners who were
accused of both a felony and a misdemeanor
is given by

$(A \cap B) = n(A) + n(B) - n(A \cup B)$
$$= 130 + 121 - 190 = 61.$$

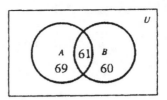

19. Let U denote the set of all customers surveyed, and let
$$A = \{ x \in U \mid x \text{ buys brand } A \}$$
$$B = \{ x \in U \mid x \text{ buys brand } B \}.$$
Then $n(U) = 120$, $n(A) = 80$,
$n(B) = 68$, and $n(A \cap B) = 42$.
Refer to the diagram at the right.
a. The number of customers who buy at least
one of these brands is
$$n(A \cup B) = 80 + 68 - 42 = 106.$$

b. The number who buy exactly one of these brands
is $n(A \cap B^C) + n(A^C \cap B) = 38 + 26 = 64$

c. The number who buy only brand A is $n(A \cap B^C) = 38$.

d. The number who buy none of these brands is $n[(A \cup B)^C] = 120 - 106 = 14$.

21. Let U denote the set of 200 investors and let
$$A = \{ x \in U \mid x \text{ uses a discount broker} \}$$
$$B = \{ x \in U \mid x \text{ uses a full-service broker} \}.$$
Refer to the diagram at the right.
a. The number of investors who use at least one
kind of broker is

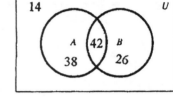

$$n(A \cup B) = n(A) + n(B) - n(A \cap B) = 120 + 126 - 64 = 182.$$

b. The number of investors who use exactly one kind of broker is
$$n(A \cap B^C) + n(A^C \cap B) = 56 + 62 = 118.$$

c. The number of investors who use only discount brokers is $n(A \cap B^C) = 56$.

d. The number of investors who don't use a broker is
$$n(A \cup B)^C = n(U) - n(A \cup B) = 200 - 182 = 18.$$

23. Let U denote the set of 200 households in the survey and let

$$A = \{x \in U \mid x \text{ owns a desktop computer}\}$$

$$B = \{x \in U \mid x \text{ owns a laptop computer}\}$$

Referring to the figure that follows, we see that the number of households that own both desktop and laptop computers is $n(A \cap B) = 200 - 120 - 10 - 40 = 30$.

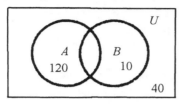

In Exercises 25 - 27, refer to the figure that follows.

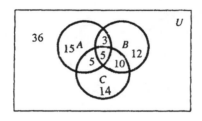

25. a. $n(A \cup B \cup C) = 64$ b. $n(A^C \cap B \cap C) = 10$

27. a. $n(A^C \cap B^C \cap C^C) = n[(A \cup B \cup C)^C] = 36$ b. $n[A^C \cap (B \cup C)] = 36$

29. Let U denote the set of all economists surveyed, and let $A = \{ x \in U \mid x$ had lowered his estimate of the consumer inflation rate$\}$

     $B = \{x \in U \mid x$ had raised his estimate of the GDP growth rate$\}$.

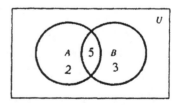

Refer to the diagram at the right. Then $n(U) = 10$, $n(A) = 7$, $n(B) = 8$, and $n(A \cap B^C) = 2$. Then the number of economists who had both lowered their estimate of the consumer inflation rate and raised their estimate of the GDP rate is given by $n(A \cap B) = 5$.

31. Let U denote the set of 100 college students who were surveyed and let

$A = \{\, x \in U \mid x$ is a student who reads *Time* magazine$\}$

$B = \{x \in U \mid x$ is a student who reads *Newsweek* magazine$\}$

and $C = \{x \in U \mid x$ is a student who reads *U.S. News and World Report* magazine$\}$

Refer to the diagram at the right.

Then $\quad n(A) = 40,\; n(B) = 30,\; n(C) = 25,\; n(A \cap B) = 15,$
$n(A \cap C) = 12,\; n(B \cap C) = 10,$ and $n(A \cap B \cap C) = 4.$

a. The number of students surveyed who read at least one magazine is
$n(A \cup B \cup C) = 17 + 11 + 4 + 8 + 6 + 7 + 9 = 62$

b. The number of students surveyed who read exactly one magazine is

$$n(A \cap B^C \cap C^C) + n(A^C \cap B \cap C^C) + n(A^C \cap B^C \cap C)$$
$$= 17 + 9 + 7 = 33.$$

c. The number of students surveyed who read exactly two magazines is

$$n(A \cap B \cap C^C) + n(A^C \cap B \cap C) + n(A \cap B^C \cap C)$$
$$= 11 + 6 + 8 = 25.$$

d. The number of students surveyed who did not read any of these magazines is

$$n(A \cup B \cup C)^C = 100 - 62 = 38.$$

33. Let U denote the set of all customers surveyed, and let
$A = \{x \in U \mid x$ buys brand $A\}$

$B = \{x \in U \mid x$ buys brand $B\}.$

$C = \{x \in U \mid x$ buys brand $C\}.$

Refer to the figure at the right. Then

$n(U) = 120,\; n(A \cap B \cap C^C) = 15,$

$n(A^C \cap B \cap C^C) = 25,$

$n(A^C \cap B^C \cap C) = 26,$

$n(A \cap B \cap C^C) = 15,\; n(A \cap B^C \cap C) = 10,$

$n(A^C \cap B \cap C) = 12,$ and $n(A \cap B \cap C) = 8.$

a. The number of customers who buy at least one of these brands is
$n(A \cup B \cup C) = 12 + 15 + 25 + 12 + 8 + 10 + 26 = 108.$

b. The number who buy labels A and B but not C is $n(A \cap B \cap C^C) = 15$

c. The number who buy brand A is $n(A) = 12 + 10 + 15 + 8 = 45$.

d. The number who buy none of these brands is
$$n[(A \cup B \cup C)^c] = 120 - 108 = 12.$$

35. Let U denote the set of 200 employees surveyed, and let
$$A = \{x \in U \mid x \text{ had investments in stock funds}\}$$
$$B = \{x \in U \mid x \text{ had investments in bond funds}\}$$
$$C = \{x \in U \mid x \text{ had investments in money market funds}\}$$

Then
$$n(U) = 200, \; n(A) = 141, \; n(B) = 91, \; n(c) = 60, \; n(A \cap B) = 47,$$
$$n(A \cap C) = 36, \; n(B \cap C) = 36, \text{ and } n(A^c \cap B^c \cap C^c) = n[(A \cup B \cup C)^c] = 5$$

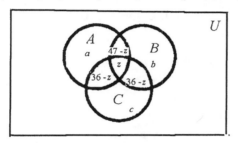

Letting $n(A \cap B \cap C) = z$ and using the fact that $n(A \cap B) = 47$, $n(A \cap C) = 36$, and $n(B \cap C) = 36$, leads to the Venn diagram shown. Next, using the fact that $n(A) = 141$, $n(B) = 91$, and $n(C) = 60$ leads to
$$a + (36 - z) + (47 - z) + z = 141$$
$$b + (47 - z) + (36 - z) + z = 91$$
$$c + (36 - z) + (36 - z) + z = 60$$
$$a + b + c + (36 - z) + (47 - z) + (36 - z) + z + 5 = 200$$

which simplifies to $a - z = 58$, $b - z = 8$, $c - z = -12$, $a + b + c - 2z = 76$.
Solving, we find $a = 80$, $b = 30$, $c = 10$, and $z = 22$. Therefore,

a. The number of employees surveyed who had invested in all three vehicles is
$n(A \cap B \cap C) = z = 22$.

b. The number who had invested in stock funds only is given by
$n(A \cap B^c \cap C^c) = a = 80$.

37. True. $n(A \cup B) = n(A) + n(B) - n(A \cap B)$.

39. True. If $A \cap B \neq \varnothing$, then $n(A \cup B) = n(A) + n(B) - n(A \cap B)$.

EXERCISES 6.3, page 357

1. By the multiplication principle, the number of rates is given by $(4)(3) = 12$.

3. By the multiplication principle, the number of ways that a blackjack hand can be dealt is $(4)(16) = 64$.

5. By the multiplication principle, she can create $(2)(4)(3) = 24$ different ensembles.

7. The number of paths is $2 \times 4 \times 3$, or 24.

9. By the multiplication principle, we see that the number of ways a health-care plan can be selected is $(10)(3)(2) = 60$.

11. $10^9 = 1,000,000,000$.

13. The number of different responses is $\underbrace{(5)\,(5)\,...\,(5)}_{50 \text{ terms}} = 5^{50}$.

15. The number of selections is given by $(2)(5)(3)$, or 30 selections.

17. The number of different selections is $(10)(10)(10)(10) - 10 = 10000 - 10 = 9990$.

19 . a. The number of license plate numbers that may be formed is
$(26)(26)(26)(10)(10)(10)$, or 17,576,000.
   b. The number of license plate numbers that may be formed is
$(10)(10)(10)(26)(26)(26) = 17,576,000$.

21. If every question is answered, there are 2^{10}, or 1024, ways. In the second case, there are 3 ways to answer each question, and so we have 3^{10}, or 59,049, ways.

23. The number of ways the first, second and third prizes can be awarded is
$(15)(14)(13) = 2730$.

25. The number of ways in which the nine symbols on the wheels can appear in the window slot is $(9)(9)(9)$, or 729. The number of ways in which the eight symbols

other than the "lucky dollar" can appear in the window slot is (8)(8)(8) or 512. Therefore, the number of ways in which the "lucky dollars" can appear in the window slot is 729 - 512, or 217.

27. True. There are 4 choices for the digit in the hundreds position, 4 choices in the tens position and 2 choices in the units position, or $4 \cdot 4 \cdot 2$, ore 32 such numbers.

EXERCISES 6.4, page 370

1. $3(5!) = 3(5)(4)(3)(2)(1) = 360.$

3. $\dfrac{5!}{2!3!} = 5(2) = 10.$

5. $P(5,5) = \dfrac{5!}{(5\text{-}5)!} = \dfrac{5!}{0!} = 120$

7. $P(5,2) = \dfrac{5!}{(5-2)!} = \dfrac{5!}{3!} = (5)(4) = 20$

9. $P(n,1) = \dfrac{n!}{(n-1)!} = n$

11. $C(6,6) = \dfrac{6!}{6!0!} = 1$

13. $C(7,4) = \dfrac{7!}{4!3!} = \dfrac{7\cdot6\cdot5}{3\cdot2} = 35$

15. $C(5,0) = \dfrac{5!}{5!0!} = 1$

17. $C(9,6) = \dfrac{9!}{3!6!} = \dfrac{9\cdot8\cdot7}{3\cdot2} = 84$

19. $C(n,2) = \dfrac{n!}{(n-2)!2!} = \dfrac{n(n-1)}{2}$

21. $P(n,n-2) = \dfrac{n!}{(n-(n-2))!} = \dfrac{n!}{(n-n+2)!} = \dfrac{n!}{2}$

23. Order is important here since the word "*glacier*" is different from "*reicalg*", so this is a permutation.

25. Order is not important here. Therefore, we are dealing with a combination. If we consider a sample of three record-o-phones of which one is defective, it does not matter whether the defective record-o-phone is the first member of our sample, the second member of our sample, or the third member of our sample. The net result is a sample of three record-o-phones of which one is defective.

27. The order is important here. Therefore, we are dealing with a permutation. Consider,

for example, 9 books on a library shelf. Each of the 9 books would have a call number, and the books would be placed in order of their call numbers; that is, a call number of 902 would come before a call number of 910.

29. The order is not important here, and consequently we are dealing with a combination. It would not matter if the hand $Q\ Q\ Q\ 5\ 5$ were dealt or the hand $5\ 5\ Q\ Q\ Q$. In each case the hand would consist of three queens and a pair.

31. The number of 4-letter permutations is $P(4,4) = \dfrac{4!}{0!} = 4 \cdot 3 \cdot 2 \cdot 1 = 24$.

33. The number of seating arrangements is $P(4,4) = \dfrac{4!}{0!} = 24$.

35. The number of different batting orders is $P(9,9) = \dfrac{9!}{0!} = 362,880$.

37. The number of different ways the 3 candidates can be selected is
$$C(12,3) = \frac{12!}{9!3!} = \frac{12 \cdot 11 \cdot 10}{3 \cdot 2 \cdot 1} = 220.$$

39. There are 10 letters in the word *ANTARCTICA*, 3*A*s, 1*N*, 2*T*s, 1*R*, 2*C*s, and 1*I*. Therefore, we use the formula for the permutation of n objects, not all distinct:
$$\frac{n!}{n_1! n_2! \cdots n_r!} = \frac{10!}{3!2!2!} = 151,200$$

41. The number of ways the 3 sites can be selected is
$$C(12,3) = \frac{12!}{9!3!} = \frac{12 \cdot 11 \cdot 10}{3 \cdot 2 \cdot 1} = 220$$

43. The number of ways in which the sample of 3 transistors can be selected is
$$C(100,3) = \frac{100!}{97!3!} = \frac{100 \cdot 99 \cdot 98}{3 \cdot 2 \cdot 1} = 161,700.$$

45. In this case order is important, as it makes a difference whether a commercial is shown first, last, or in between. The number of ways that the director can schedule the commercials is given by $P(6,6) = 6! = 720$.

47. The inquiries can be directed in

$$P(12,6) = \frac{12!}{6!} = 12 \cdot 11 \cdot 10 \cdot 9 \cdot 8 \cdot 7 = 665,280, \quad \text{or } 665,280 \text{ ways.}$$

49. a. The ten books can be arranged in

$$P(10,10) = 10! = 3,628,800 \text{ ways.}$$

b. If books on the same subject are placed together, then they can be arranged on the shelf

$$P(3,3) \times P(4,4) \times P(3,3) \times P(3,3) = 5184 \text{ ways.}$$

Here we have computed the number of ways the mathematics books can be arranged times the number of ways the social science books can be arranged times the number of ways the biology books can be arranged times the number of ways the 3 sets of books can be arranged.

51. Notice that order is certainly important here.

a. The number of ways that the 20 featured items can be arranged is given by

$$P(20,20) = 20! = 2.43 \times 10^{18}.$$

b. If items from the same department must appear in the same row, then the number of ways they can be arranged on the page is

Number of ways of arranging the rows	x	Number of ways of arranging the items in each of the 5 rows
$P(5,5)$	•	$P(4,4) \times P(4,4) \times P(4,4) \times P(4,4) \times P(4,4)$

$$= 5! \times (4!)^5 = 955,514,880.$$

53. The number of ways is given by

$$2\ \{C(2,2) + [C(3,2) - C(2,2)]\} = 2[1 + (3 - 1)] = 2 \times 3 = 6$$

(number of players)[number of ways to win in exactly 2 sets + number of ways to win in exactly 3 sets]

55. The number of ways the measure can be passed is

$$C(3,3) \times [C(8,6) + C(8,7) + C(8,8)] = 37.$$

Here three of the three permanent members must vote for passage of the bill and this can be done in $C(3,3) = 1$ way. Of the 8 nonpermanent members who are voting 6 can vote for passage of the bill, or 7 can vote for passage, or 8 can vote for passage. Therefore, there are

$$C(8,6) + C(8,7) + C(8,8) = 37 \text{ ways}$$

that the nonpermanent members can vote to ensure passage of the measure. This gives $1 \times 37 = 37$ ways that the members can vote so that the bill is passed.

57. a. If no preference is given to any student, then the number of ways of awarding the 3 teaching assistantships is $C(12,3) = \dfrac{12!}{3!9!} = 220.$

b. If it is stipulated that one particular student receive one of the assistantships, then the remaining two assistantships must be awarded to two of the remaining 11 students. Thus, the number of ways is $C(11,2) = \dfrac{11!}{2!9!} = 55.$

c. If at least one woman is to be awarded one of the assistantships, and the group of students consists of seven men and five women, then the number of ways the assistantships can be awarded is given by
$$C(5,1) \times C(7,2) \; + \; C(5,2) \times C(7,1) \; + \; C(5,3)$$
$$= \frac{5!}{4!1!} \cdot \frac{7!}{5!2!} + \frac{5!}{3!2!} \cdot \frac{7!}{6!1!} + \frac{5!}{3!2!} = 105 + 70 + 10 = 185.$$

59. The number of ways of awarding the 7 contracts to 3 different firms is given by
$$P(7,3) = \frac{7!}{4!} = 210.$$

The number of ways of awarding the 7 contracts to 2 different firms is

$$C(7,2) \times P(3,2) = 126. \qquad \text{(First pick the two firms, and}$$
$$\text{then award the 7 contracts.)}$$

Therefore, the number of ways the contracts can be awarded if no firm is to receive more than 2 contracts is given by $210 + 126 = 336.$

61. The number of different curricula that are available for the student's consideration is given by

$$C(5,1) \times C(3,1) \times C(6,2) \times C(4,1) \; + \; C(5,1) \times C(3,1) \times C(6,2) \times C(3,1)$$
$$= \frac{5!}{4!1!} \cdot \frac{3!}{2!1!} \cdot \frac{6!}{4!2!} \cdot \frac{4!}{3!1!} + \frac{5!}{4!1!} \cdot \frac{3!}{2!1!} \cdot \frac{6!}{4!2!} \cdot \frac{3!}{2!1!}$$
$$= (5)(3)(15)(4) + (5)(3)(15)(3) = 900 + 675 = 1575.$$

63. The number of ways is given by $C(10,2) + C(10,1) + C(10,0) = 45 + 10 + 1 = 56$
or $\qquad C(10,8) + C(10,9) + C(10,10) = 56.$

65. The number of ways of dealing a straight flush (5 cards in sequence in the same suit is given by

the number of ways of selecting   ×   the number of ways
   5 cards in sequence in the same suit   of selecting a suit
$$10 \quad \bullet \quad C(4,1) = \quad 40.$$

67. The number of ways of dealing a flush (5 cards in one suit that are not all in

sequence) is given by

the number of ways of selecting   -   the number of ways of selecting
      5 cards in one suit   5 cards in one suit in sequence
$$4C(13,5) \quad - \quad 4(10)$$
$$= 5148 - 40 = 5108.$$

69. The number of ways of dealing a full house (3 of a kind and a pair) is given by

the number of ways of picking   ×   the number of ways of picking a pair
   3 of a kind from a given rank   from the 12 remaining ranks

$$13C(4,3) \bullet 12C(4,2)$$
$$= 13(4) \quad \bullet \quad (12)(6) = 3744.$$

71. The bus will travel a total of 6 blocks. Each route must include 2 blocks running north and south and 4 blocks running east and west. To compute the total number of possible routes, it suffices to compute the number of ways the 2 blocks running north and south can be selected from the six blocks. Thus,
$$C(6,2) = \frac{6!}{2!4!} = 15.$$

73. The number of ways that the quorum can be formed is given by
$$C(12,6) + C(12,7) + C(12,8) + C(12,9) + C(12,10) + C(12,11) + C(12,12)$$
$$= \frac{12!}{6!6!} + \frac{12!}{7!5!} + \frac{12!}{8!4!} + \frac{12!}{9!3!} + \frac{12!}{10!2!} + \frac{12!}{11!1!} + \frac{12!}{12!0!}$$
$$= 924 + 792 + 495 + 220 + 66 + 12 + 1 = 2510.$$

75. Using the formula given in Exercise 74, we see that the number of ways of seating the 5 commentators at a round table is
$$(5 - 1)! = 4! = 24.$$

77. The number of possible corner points is $C(8,3) = \dfrac{8!}{5!3!} = 56.$

79. True.

81. True. $C(n,r) = \dfrac{n!}{(n-r)!r!}$ and $C(n,n-r) = \dfrac{n!}{[n-(n-r)]!(n-r)!} = \dfrac{n!}{r!(n-r)!}$.

So, $C(n,r) = C(n,n-r)$.

USING TECHNOLOGY EXERCISES 6.4, page 373

1. $1.307674368 \times 10^{12}$ 3. $2.56094948229 \times 10^{16}$

5. 674,274,182,400   7. 133,784,560   9. 4,656,960

11. Using the multiplication principle, the number of 10-question exams she can set is given by $C(25,3) \times C(40,5) \times C(30,2) = 658,337,004,000$.

CHAPTER 6, REVIEW EXERCISES, page 376

1. {3}. The set consists of all solutions to the equation $3x - 2 = 7$.

3. {4,6,8,10}   5. Yes.   7. Yes.

9.   11.

$A \cup (B \cap C)$

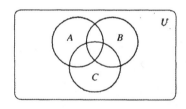

$A^c \cap B^c \cap C^c$

13. $A \cup (B \cup C) = \{a,b\} \cup [\{b,c,d\} \cup \{a,d,e\}]$
$= \{a,b\} \cup \{a,b,c,d,e\} = \{a,b,c,d,e\}$.
$(A \cup B) \cup C = [\{a,b\} \cup \{b,c,d\}] \cup \{a,d,e\}$
$= \{a,b,c,d\} \cup \{a,d,e\} = \{a,b,c,d,e\}$.

15. $A \cap (B \cup C) = \{a,b\} \cap [\{b,c,d\} \cup \{a,d,e\}] = \{a,b\} \cap \{a,b,c,d,e\} = \{a,b\}$.
$(A \cap B) \cup (A \cap C) = [\{a,b\} \cap \{b,c,d\}] \cup [\{a,b\} \cap \{a,d,e\} = \{b\} \cup \{a\} = \{a,b\}$.

17. The set of all participants in a consumer behavior survey who both avoided buying a product because it is not recyclable and boycotted a company's products because of its record on the environment.

19. The set of all participants in a consumer behavior survey who both did not use cloth diapers rather than disposable diapers and voluntarily recycled their garbage.

In Exercises 21-25, refer to the following Venn diagram.

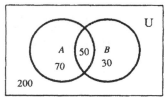

21. $n(A \cup B) = n(A) + n(B) - n(A \cap B) = 120 + 80 - 50 = 150$.

23. $n(B^c) = n(U) - n(B) = 350 - 80 = 270$.

25. $n(A \cap B^c) = n(A) - n(A \cap B) = 120 - 50 = 70$.

27. $C(20,18) = \dfrac{20!}{18!2!} = 190$.

29. $C(5,3) \cdot P(4,2) = \dfrac{5!}{3!2!} \cdot \dfrac{4!}{2!} = 10 \cdot 12 = 120$.

31. Let U denote the set of 5 major cards, and let
    $A = \{ x \in U \mid x \text{ offered cash advances} \}$
    $B = \{ x \in U \mid x \text{ offered extended payments for all goods and services purchased} \}$
    $C = \{ x \in U \mid x \text{ required an annual fee that was less than \$35} \}$

Thus, $n(A) = 3$, $n(B) = 3$, $n(C) = 2$
     $n(A \cap B) = 2$, $n(B \cap C) = 1$
and $n(A \cap B \cap C) = 0$
Using the Venn diagram on the right, we have
     $x + y + 2 = 3$
     $y + 2 = 2$
Solving the system, we find $x = 1$, and $y = 0$.
Therefore, the number of cards that offer cash advances and have an annual fee that

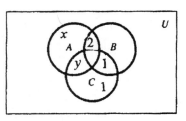

is less than \$35 is given by $n(A \cap C) = y = 0$.

33. The number of ways the compact discs can be arranged on a shelf is
$$P(6,6) = 6! = 720 \text{ ways.}$$
.

35. a. Since there is repetition of the letters C, I, and N, we use the formula for the permutation of n objects, not all distinct, with $n = 10$, $n_1 = 2$, $n_2 = 3$, and $n_3 = 3$. Then the number of permutations that can be formed is given by
$$\frac{10!}{2!3!3!} = 50{,}400.$$
b. Here, again, we use the formula for the permutation of n objects, not all distinct, this time with $n = 8$, $n_1 = 2$, $n_2 = 2$, and $n_3 = 2$. Then the number of permutations is given by $\dfrac{8!}{2!2!2!} = 5040$.

37. Let U denote the set comprising Halina's clients, and let
$$A = \{x \in U \,|\, x \text{ owns stocks}\}$$
$$B = \{x \in U \,|\, x \text{ owns bonds}\}$$
$$C = \{x \in U \,|\, x \text{ owns mutual funds}\}$$
Then $n(A) = 300$, $n(B) = 180$, $n(C) = 160$, $n(A \cap B) = 110$, $n(A \cap C) = 120$, $n(B \cap C) = 90$. Let $n(A \cap B \cap C) = z$. Then we are lead to the following Venn diagram.

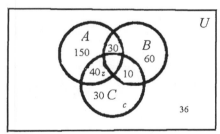

Next, using the fact that $n(A) = 300$, $n(B) = 180$, and $n(C) = 160$ leads to
$$a + (110 - z) + (120 - z) + z = 300$$
$$b + (110 - z) + (90 - z) + z = 180$$
$$c + (120 - z) + (90 - z) + z = 160$$
$$a + b + c + (110 - z) + (120 - z) + (90 - z) + z = 400$$
which simplifies to $a - z = 70$, $b - z = -20$, $c - z = -50$, and $a + b + c - 2z = 80$.

solving, we find $a = 150$, $b = 60$, $c = 30$ and $z = 80$. therefore, the number who own stocks, bonds and mutual funds is $n(A \cap B \cap C) = z = 80$.

39. The number of possible outcomes is $(6)(4)(5)(6) = 720$.

41. a. The number of ways the 7 students can be assigned to seats is $P(7,7) = 7! = 5040$.

    b. The number of ways 2 specified students can be seated next to each other is
       $2(6) = 12$.
    (Think of seven numbered seats. Then the students can be seated in seats 1-2, or 2-3, or 3-4, or 4-5, or 5-6, or 6-7. Since there are 6 such possibilities and the pair of students can be seated in 2 ways, we conclude that there are 2(6) possible arrangements.)
    Then the remaining 5 students can be seated in $P(5,5) = 5!$ ways. Therefore, the number of ways the 7 students can be seated if two specified students sit next to each other is $2(6)5! = 1440$.
    Finally, the number of ways the students can be seated if the two students do not sit next to each other is $P(7,7) - 2(6)5! = 5040 - 1440 = 3600$.

43. a. The number of samples that can be selected is $C(15,4) = \dfrac{15!}{4!11!} = 1365$.

    b. There are $C(10,4) = \dfrac{10!}{4!6!} = 210$ ways of selecting 4 balls none of which are white.
    Therefore, there are 1365 - 210, or 1155 ways of selecting 4 balls of which at least one is white.

CHAPTER 7

1. $E \cup F = \{a,b,d,f\}$; $E \cap F = \{a\}$.

3. $F^C = \{b,c,e\}$; $E \cap G^C = \{a,b\} \cap \{a,d,f\} = \{a\}$.

5. Since $E \cap F = \{a\}$ is not a null set, we conclude that E and F are not mutually exclusive.

7. $E \cup F \cup G = \{2,4,6\} \cup \{1,3,5\} \cup \{5,6\} = \{1,2,3,4,5,6\}$.

9. $(E \cup F \cup G)^C = \{1,2,3,4,5,6\}^C = \varnothing$.

11. Yes, $E \cap F = \varnothing$; that is, E and F do not contain any common elements.

13. $E^c = \{2,4,6\}^c = \{1,3,5\} = F$ and so E and F are complementary.

15. $E \cup F$ 17. G^C 19. $(E \cup F \cup G)^C$

21. a. Refer to Example 4, page 383.
$$E = \{(2,1),(3,1),(4,1),(5,1),(6,1),(3,2),(4,2),(5,2),(6,2),$$
$$(4,3),(5,3),(6,3),(5,4),(6,4),(6,5)\}$$
    b. $E = \{(1,2),(2,4),(3,6)\}$

23. $\varnothing$, $\{a\}$, $\{b\}$, $\{c\}$, $\{a,b\}$, $\{a,c\}$, $\{b,c\}$, $\{a,b,c\}$.

25. a. $S = \{R,B\}$ b. $\varnothing$, $\{B\}$, $\{R\}$, $\{B,R\}$

27. a. $S = \{(H,1), (H,2), (H,3), (H,4), (H,5), (H,6), (T,1), (T,2),$
$$(T,3), (T,4), (T,5), (T,6)\}$$
    b. $E = \{(H,2), (H,4), (H,6)\}$

29. a. Here $S = \{1,2,3,4,5,6\}$, $E = \{2\}$, $F = \{2,4,6\}$. Since $E \cap F = \{2\} \neq \emptyset$, we conclude that E and F are not mutually exclusive.
    b. $E^c = \{1,3,4,5,6\} \neq F$ and so E and F are not complementary.

31. $S = \{(d,d,d), (d,d,n), (d,n,d), (n,d,d), (d,n,n), (n,d,n), (n,n,d), (n,n,n)\}$

33. a. $\{ABC, ABD, ABE, ACD, ACE, ADE, BCD, BCE, BDE, CDE\}$;
    b. 6   c. 3   d. 6

35. a. E^C b. $E^C \cap F^C$ c. $E \cup F$ d. $(E \cap F^C) \cup (E^C \cap F)$

37. a. $S = \{t \mid t > 0\}$ b. $E = \{t \mid 0 < t \leq 2\}$ c. $F = \{t \mid t > 2\}$

39. a. $S = \{0,1,2,3,...,10\}$ b. $E = \{0,1,2,3\}$ c. $F = \{5,6,7,8,9,10\}$

41. a. $S = \{0,1,2,...,20\}$ b. $E = \{0,1,2,...,9\}$ c. $F = \{20\}$

43. Let S denote the sample space of the experiment that is the set of 52 cards. Then $E = \{x \in S \mid x \text{ is an ace}\}$ and $F = \{x \in S \mid x \text{ is a spade}\}$ and $E \cap F = \{x \in S \mid x \text{ is the ace of spades}\}$. Now $n(E) = 4$, $n(F) = 13$, and $n(E \cap F) = 1$.

    Also, $E \cup F = \{x \in S \mid x \text{ is an ace or a spade}\}$ and $n(E \cup F) = 16$, and
    $$n(E) + n(F) - n(E \cap F) = 4 + 13 - 1 = 16 = n(E \cup F).$$

45. $E^C \cap F^C = (E \cup F)^C$ by DeMorgan's Law. Since $(E \cup F) \cap (E \cup F)^C = \emptyset$, they are mutually exclusive.

47. False. Let $E = \{1, 2, 3\}$, $F = \{4, 5, 6\}$, and $G = \{4, 5\}$. Then $E \cap F = \emptyset$ and $E \cap G = \emptyset$, but $F \cap G = \{4,5\} \neq \emptyset$.

EXERCISES 7.2, page 394

1. $\{(H,H)\}$, $\{(H,T)\}$, $\{(T,H)\}$, $\{(T,T)\}$.

3. $\{(D,m)\}$, $\{(D,f)\}$, $\{(R,m)\}$, $\{(R,f)\}$, $\{(I,m)\}$, $\{(I,f)\}$

5. $\{(1,i)\}, \{(1,d)\}, \{(1,s)\}, \{(2,i)\}, \{(2,d)\}, \{(2,s)\}, \ldots, \{(5,i)\}, \{(5,d)\}, \{(5,s)\}$

7. $\{(A, Rh^+)\}, \{(A, Rh^-)\}, \{(B, Rh^+)\}, \{B, Rh^-)\}, \{(AB, Rh^+)\}, \{(AB, Rh^-)\},$
    $\{(O, Rh^+)\}, \{(O, Rh^-)\}$

9. The probability distribution associated with this data is

Grade	A	B	C	D	F
Probability	0.10	0.25	0.45	0.15	0.05

11. a. $S = \{(0 < x \le 200), (200 < x \le 400), (400 < x \le 600), (600 < x \le 800),$
    $(800 < x \le 1000), (x > 1000)\}$

    b.

Number of cars (x)	Probability
$0 < x \le 200$	0.075
$200 < x \le 400$	0.1
$400 < x \le 600$	0.175
$600 < x \le 800$	0.35
$800 < x \le 1000$	0.225
$x > 1000$	0.075

13. The probability distribution associated with this data is

Rating	A	B	C	D	F
Probability	0.026	0.199	0.570	0.193	0.012

15. The probability distribution is

Number of figures produced (in dozens)	30	31	32	33	34	35	36
Probability	0.125	0	0.1875	0.25	0.1875	0.125	0.125

17. The probability is $\dfrac{84,000,000}{179,000,000} \approx 0.469$.

19. a. The probability that a person killed by lightning is a male is $\dfrac{376}{439} \approx 0.856$.

    b. The probability that a person killed by lightning is a female is
$$\dfrac{439-376}{439} = \dfrac{63}{439} \approx 0.144.$$

21. The probability that the retailer uses electronic tags as antitheft devices is
$$\dfrac{81}{179} \approx 0.46.$$

23. a. $P(D) = \dfrac{13}{52} = \dfrac{1}{4}$ b. $P(B) = \dfrac{26}{52} = \dfrac{1}{2}$ c. $P(A) = \dfrac{4}{52} = \dfrac{1}{13}$

25. The probability of arriving at the traffic light when it is red is
$$\dfrac{30}{30+5+45} = \dfrac{30}{80} = 0.375.$$

27. The required event is $E = \{2,4,\dots, 36\}$. So
$$P(E) = \dfrac{1}{38} + \dfrac{1}{38} + \cdots + \dfrac{1}{38} \qquad \text{(eighteen terms)}$$
$$= \dfrac{18}{38}, \text{ or } \dfrac{9}{19}.$$

29. The probability is
$$P(600 < x \le 800) + P(800 < x \le 1000) + P(x > 1000)$$
$$= 0.35 + 0.225 + 0.075 = 0.65.$$

31. There are two ways of getting a 7, one die showing a 3 and the other die showing a 4, and vice versa.

33. No, the outcomes are not equally likely.

35. No. Since the coin is weighted, the outcomes are not equally likely.

37. a. $P(A) = P(s_1) + P(s_3) = \dfrac{1}{12} + \dfrac{1}{12} = \dfrac{1}{6}$

    b. $P(B) = P(s_2) + P(s_4) + P(s_5) + P(s_6) = \dfrac{1}{4} + \dfrac{1}{6} + \dfrac{1}{3} + \dfrac{1}{12} = \dfrac{5}{6}$ c. $P(C) = 1$.

39. Let G denote a female birth and let B denote a male birth. Then the eight equally likely outcomes of this experiment are
$$GGG \ GGB \ GBG \ BGG \ BGB \ BBG \ GBB \ BBB.$$
    a. The event that there are two girls and a boy in the family is
$$E = \{GGB, GBG, BGG\}.$$
Since there are three favorable outcomes, $P(E) = 3/8$.
    b. The event that the oldest child is a girl is
$$F = \{GGG, GGB, GBG, GBB\}.$$
Since there are 4 favorable outcomes, $P(F) = 1/2$.
    c. The event that the oldest child is a girl and the youngest child is a boy is
$$G = \{GGB, GBB\}.$$
Since there are two favorable outcomes, $P(F) = 1/4$.

41. The probability that the primary cause of the crash was due to pilot error or bad weather is given by $\dfrac{327 + 22}{327 + 49 + 14 + 22 + 19 + 15} = \dfrac{349}{446} \approx 0.7825$.

43. False. $P(s_1) + P(s_2) + \cdots + P(s_n) = 1$

EXERCISES 7.3, page 403

1. Refer to Example 4, page 383. Let E denote the event of interest. Then
$$P(E) = \dfrac{18}{36} = \dfrac{1}{2}$$

3. Refer to Example 4, page 383. The event of interest is $E = \{1,1\}$, and $P(E) = 1/36$.

5. Let E denote the event of interest. Then $E = \{(6,2),(6,1),(1,6),(2,6)\}$
    and $P(E) = \dfrac{4}{36} = \dfrac{1}{9}$.

7. Let E denote the event that the card drawn is a king, and let F denote the event that the card drawn is a diamond. Then the required probability is $P(E \cap F) = \dfrac{1}{52}$.

9. Let E denote the event that a face card is drawn. Then $P(E) = \dfrac{12}{52} = \dfrac{3}{13}$.

11. Let E denote the event that an ace is drawn. Then $P(E) = 1/13$. Then E^c is the event that an ace is not drawn and $P(E^C) = 1 - P(E) = \dfrac{12}{13}$.

13. Let E denote the event that a ticket holder will win first prize, then
$$P(E) = \dfrac{1}{500} = 0.002,$$
and the probability of the event that a ticket holder will not win first prize is
$P(E^c) = 1 - 0.002 = 0.998$.

15. Property 2 of the laws of probability is violated. The sum of the probabilities must add up to 1. In this case $P(S) = 1.1$, which is not possible.

17. The five events are not mutually exclusive; the probability of winning at least one purse is
$$1 - \text{probability of losing all 5 times} = 1 - \dfrac{9^5}{10^5} = 1 - 0.5905 = 0.4095.$$

19. The two events are not mutually exclusive; hence, the probability of the given event is $\dfrac{1}{6} + \dfrac{1}{6} - \dfrac{1}{36} = \dfrac{11}{36}$.

21. $E^C \cap F^C = \{c, d, e\} \cap \{a, b, e\} = \{e\} \neq \varnothing$.

23. Let G denote the event that a customer purchases a pair of glasses and let C denote the event that the customer purchases a pair of contact lenses. Then
$$P(G \cup C)^C \neq 1 - P(G) - P(C).$$
Mr. Owens has not considered the case in which the customer buy both glasses and contact lenses.

25. a. $P(E \cap F) = 0$ since E and F are mutually exclusive.
    b. $P(E \cup F) = P(E) + P(F) - P(E \cap F) = 0.2 + 0.5 = 0.7$.
    c. $P(E^c) = 1 - P(E) = 1 - 0.2 = 0.8$.
    d. $P(E^C \cap F^C) = P[(E \cup F)^C] = 1 - P(E \cup F) = 1 - 0.7 = 0.3$.

27. a. $P(A) = P(s_1) + P(s_2) = \dfrac{1}{8} + \dfrac{3}{8} = \dfrac{1}{2}$; $P(B) = P(s_1) + P(s_3) = \dfrac{1}{8} + \dfrac{1}{4} = \dfrac{3}{8}$

    b. $P(A^C) = 1 - P(A) = 1 - \dfrac{1}{2} = \dfrac{1}{2}$; $P(B^C) = 1 - P(B) = 1 - \dfrac{3}{8} = \dfrac{5}{8}$

    c. $P(A \cap B) = P(s_1) = \dfrac{1}{8}$

    d. $P(A \cup B) = P(A) + P(B) - P(A \cap B) = \dfrac{1}{2} + \dfrac{3}{8} - \dfrac{1}{8} = \dfrac{3}{4}$

29. Referring to the following diagram we see that

    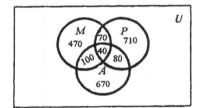

    P: lack of parental support
    M: malnutrition
    A: abused or neglected

    the probability that a teacher selected at random from this group said that lack of parental support is the only problem hampering a student's schooling is

    $$\frac{710}{2140} = 0.33.$$

31. Let E and F denote the events that the person surveyed learned of the products from *Good Housekeeping* and *The Ladies Home Journal*, respectively. Then

    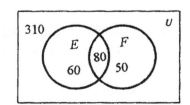

    $$P(E) = \frac{140}{500} = \frac{7}{25}, \quad P(F) = \frac{130}{500} = \frac{13}{50}$$

    and $P(E \cap F) = \dfrac{80}{500} = 0.16$

a. $P(E \cap F) = \dfrac{80}{500} = 0.16$ b. $P(E \cup F) = \dfrac{14}{50} + \dfrac{13}{50} - \dfrac{8}{50} = \dfrac{19}{50} = 0.38$

c. $P(E \cap F^C) + P(E^C \cap F) = \dfrac{60}{500} + \dfrac{50}{500} = \dfrac{110}{500} = 0.22$.

33. Let $A = \{t \mid t < 3\}$, $B = \{t \mid t \le 4\}$, $C = \{t \mid t > 5\}$

a. $D = \{t \mid t \le 5\}$ and $P(D) = 1 - P(C) = 1 - 0.1 = 0.9$.

b. $E = \{t \mid t > 4\}$ and $P(E) = 1 - P(B) = 1 - 0.6 = 0.4$.

c. $F = \{t \mid 3 \le t \le 4\}$ and $P(F) = P(A^C \cap B) = 0.4$.

35. a. The probability that the participant favors tougher gun-control laws is $\dfrac{150}{250} = 0.6$.

b. The probability that the participant owns a handgun is $\dfrac{58 + 25}{250} = 0.332$.

c. The probability that the participant owns a handgun but not a rifle is

$\dfrac{58}{250} = 0.232$.

d. The probability that the participant favors tougher gun-control laws and does not own a handgun is $\dfrac{12 + 138}{250} = 0.6$.

37. The probability that Bill will fail to solve the problem is $1 - p_1$, and the probability that Mike will fail to solve the problem is $1 - p_2$. Therefore, the probability that both Bill and Mike will fail to solve the problem is $(1 - p_1)(1 - p_2)$. So, the probability that at least one of them will solve the problem is

$1 - (1 - p_1)(1 - p_2) = 1 - (1 - p_2 - p_1 + p_1 p_2) = p_1 + p_2 - p_1 p_2$

39. True. Write $B = A \cup (B - A)$. Since A and $B - A$ are mutually exclusive, we have
$P(B) = P(A) + P(B - A)$.

Since $P(B) = 0$, we have $P(A) + P(B - A) = 0$. If $P(A) > 0$, then $P(B - A) < 0$ and this is not possible. Therefore, $P(A) = 0$.

41. False. Take $E_1 = \{1, 2\}$ and $E_2 = \{2, 3\}$ where $S = \{1, 2, 3\}$. Then
$P(E_1) = \frac{2}{3}$, $P(E_2) = \frac{2}{3}$, but $P(E_1 \cup E_2) = P(S) = 1$.

EXERCISES 7.4, page 411

1. Let E denote the event that the coin lands heads all five times. Then
$$P(E) = \frac{1}{2^5} = \frac{1}{32}.$$

3. Let E denote the event that the coin lands tails all 5 times, then
$$P(E^c) = 1 - P(E) = 1 - \frac{1}{32} = \frac{31}{32},$$
where E^c is the event that the coin lands heads at least once.

5. $P(E) = \dfrac{13 \cdot C(4,2)}{C(52,2)} = \dfrac{78}{1326} \approx 0.0588$.

7. $P(E) = \dfrac{C(26,2)}{C(52,2)} = \dfrac{325}{1326} \approx 0.2451$.

9. The probability of the event that two of the balls will be white and two will be blue
is $P(E) = \dfrac{n(E)}{n(S)} = \dfrac{C(3,2) \cdot C(5,2)}{C(8,4)} = \dfrac{(3)(10)}{70} = \dfrac{3}{7}$.

11. The probability of the event that exactly three of the balls are blue is
$$P(E) = \frac{n(E)}{n(S)} = \frac{C(5,3)C(3,1)}{C(8,4)} = \frac{30}{70} = \frac{3}{7}.$$

13. $\qquad P(E) = \dfrac{C(3,2) \cdot C(1,1)}{8} = \dfrac{3}{8}$.

15. $\qquad P(E) = \dfrac{C(3,3)}{8} = \dfrac{1}{8}$.

17. The number of elements in the sample space is 2^{10}. There are $C(10,6) = \dfrac{10!}{6!4!}$,
or 210 ways of answering exactly six questions correctly. Therefore, the required
probability is $\dfrac{210}{2^{10}} = \dfrac{210}{1024} \approx 0.205$.

19. a. Let E denote the event that both of the bulbs are defective. Then

$$P(E) = \frac{C(4,2)}{C(24,2)} = \frac{\frac{4!}{2!2!}}{\frac{24!}{22!2!}} = \frac{4 \cdot 3}{24 \cdot 23} = \frac{1}{46} \approx 0.022.$$

b. Let F denote the event that none of the bulbs are defective. Then

$$P(F) = \frac{C(20,2)}{C(24,2)} = \frac{20!}{18!2!} \cdot \frac{22!2!}{24!} = \frac{20}{24} \cdot \frac{19}{23} = 0.6884.$$

Therefore, the probability that at least one of the light bulbs is defective is given by
$1 - P(F) = 1 - 0.6884 = 0.3116.$

21. a. The probability that both of the cartridges are defective is

$$P(E) = \frac{C(6,2)}{C(80,2)} = \frac{15}{3160} = 0.0047.$$

b. Let F denote the event that none of the cartridges are defective. Then

$$P(F) = \frac{C(74,2)}{C(80,2)} = \frac{2701}{3160} = 0.855,$$

and $P(F^c) = 1 - P(F) = 1 - 0.855 = 0.145$ is the probability that at least 1 of the cartridges is defective.

23. a. The probability that Mary's name will be selected is $P(E) = \frac{12}{100} = 0.12$;

The probability that both Mary's and John's names will be selected is

$$P(F) = \frac{C(98,10)}{C(100,12)} = \frac{\frac{98!}{88!10!}}{\frac{100!}{88!12!}} = \frac{12 \cdot 11}{100 \cdot 99} \approx 0.013.$$

b. The probability that Mary's name will be selected is $P(M) = \frac{6}{40} = 0.15$.

The probability that both Mary's and John's names will be selected is

$$P(M) \cdot P(J) = \frac{6}{60} \cdot \frac{6}{40} = \frac{36}{2400} = 0.015.$$

25. The probability is given by

$$\frac{C(12,8) \cdot C(8,2)}{C(20,10)} + \frac{C(12,9)C(8,1)}{C(20,10)} + \frac{C(12,10)}{C(20,10)}$$

$$= \frac{(28)(495) + (220)(8) + 66}{184{,}756} \approx 0.085$$

27. a. The probability that he will select brand B is
$$\frac{C(4,2)}{C(5,3)} = \frac{6}{10} = \frac{3}{5}. \qquad (\frac{\text{the number of selections that include brand } B}{\text{the number of possible selections}})$$

b. The probability that he will select brands B and C is
$$\frac{C(3,1)}{C(5,3)} = 0.3$$

c. The probability that he will select at least one of the two brands, B and C is
$$1 - \frac{C(3,3)}{C(5,3)} = 0.9. \qquad \text{(1 - probability that he does not select brands } B \text{ and } C.)$$

29. The probability that the three "Lucky Dollar" symbols will appear in the window of the slot machine is
$$P(E) = \frac{n(E)}{n(S)} = \frac{(1)(1)(1)}{C(9,1)C(9,1)C(9,1)} = \frac{1}{729}.$$

31. The probability of a ticket holder having all four digits in exact order is
$$\frac{1}{C(10,1) \cdot C(10,1) \cdot C(10,1) \cdot C(10,1)} = \frac{1}{10{,}000} = 0.0001.$$

33. The probability of a ticket holder having a specified digit in exact order is
$$\frac{C(1,1)C(10,1)C(10,1)C(10,1)}{10^4} = 0.10.$$

35. The number of ways of selecting a 5-card hand from 52 cards is given by
$$C(52,5) = 2{,}598{,}960.$$
The number of straight flushes that can be dealt in each suit is 10, so there are 4(10) possible straight flushes. Therefore, the probability of being dealt a straight flush is
$$\frac{4(10)}{C(52,5)} = \frac{40}{2{,}598{,}960} = 0.0000154.$$

37. The number of ways of being dealt a flush in one suit is $C(13,5)$, and, since there are four suits, the number of ways of being dealt a flush is $4 \cdot C(13,5)$. Since we wish to exclude the hands that are straight flushes we subtract the number of possible straight flushes from $4 \cdot C(13,5)$. Therefore, the probability of being drawn a flush, but not a straight flush, is

$$\frac{4 \cdot C(13,5) - 40}{C(52,5)} = \frac{5108}{2,598,960} = 0.0019654 .$$

39. The total number of ways to select three cards of one rank is $13 \cdot C(4,3)$. The remaining two cards must form a pair of another rank and there are
$$12 \cdot C(4,2)$$
ways of selecting these pairs. Next, the total number of ways to be dealt a full house is $13 \cdot C(4,3) \cdot 12 \cdot C(4,2) = 3744$.

Hence, the probability of being dealt a full house is $\dfrac{3,744}{2,598,960} \approx 0.0014406 .$

41. Let E denote the event that in a group of 5, no two will have the same sign. Then
$$P(E) = \frac{12 \cdot 11 \cdot 10 \cdot 9 \cdot 8}{12^5} \approx 0.381944.$$
Therefore, the probability that at least two will have the same sign is given by
$$1 - P(E) = 1 - 0.381944 \approx 0.618.$$

    b. $P(\text{no Aries}) = \dfrac{11 \cdot 11 \cdot 11 \cdot 11 \cdot 11}{12^5} \approx 0.647228.$

$$P(1 \text{ Aries}) = \frac{C(5,1) \cdot (1)(11)(11)(11)(11)}{12^5} \approx 0.2941945 .$$

Therefore, the probability that at least two will have the sign Aries is given by
$$1 - [P(\text{no Aries}) + P(1 \text{ Aries})] = 1 - 0.9414225 \approx 0.059.$$

43. Referring to the table on page 410, we see that in a group of 50 people, the probability that none of the people will have the same birthday is $1 - 0.970 = 0.03$.

EXERCISES 7.5, page 424

1.   a. $P(A|B) = \dfrac{P(A \cap B)}{P(B)} = \dfrac{0.2}{0.5} = \dfrac{2}{5} .$ b. $P(B|A) = \dfrac{P(A \cap B)}{P(A)} = \dfrac{0.2}{0.6} = \dfrac{1}{3} .$

3. $P(A \cap B) = P(A)P(B|A) = (0.6)(0.5) = 0.3.$

5. $P(A) \cdot P(B) = (0.3)(0.6) = 0.18 = P(A \cap B)$. Therefore the events are independent.

7. $P(A \cap B) = P(A) + P(B) - P(A \cup B) = 0.5 + 0.7 - 0.85 = 0.35 = P(A) \cdot P(B)$
   so they are independent events.

9. a. $P(A \cap B) = P(A)P(B) = (0.4)(0.6) = 0.24$.
   b. $P(A \cup B) = P(A) + P(B) - P(A \cap B) = 0.4 + 0.6 - 0.24 = 0.76$.

11. a. $P(A) = 0.5$　　　　b. $P(E|A) = 0.4$
    c. $P(A \cap E) = P(A)P(E|A) = (0.5)(0.4) = 0.2$
    d. $P(E) = (0.5)(0.4) + (0.5)(0.3) = 0.35$.
    e. No. $P(A \cap E) \neq P(A) \cdot P(E) = (0.5)(0.35)$
    f. A and E are not independent events.

13.

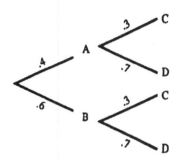

    a. $P(A) = 0.4$　　b. $P(C|A) = 0.3$　c. $P(A \cap C) = P(A)P(C|A) = (0.4)(0.3) = 0.12$
    d. $P(C) = (0.4)(0.3) + (0.6)(0.3) = 0.3$ e. Yes. $P(A \cap C) = 0.12 = P(A)P(C)$ f. Yes.

15. a. Refer to Figure 7.15, page 416. Here $E = \{(5,1),(5,2),(5,3),(5,4),(5,5),(5,6)\}$
    and $F = \{(6,4),(5,5),(4,6)\}$. So $P(F) = \frac{3}{36} = \frac{1}{12}$.
    b. $P(E \cap F) = \frac{1}{36}$ since $E \cap F = \{(5,5)\}$.　c. $P(F|E) = \frac{1}{6}$.　　　　d. $P(E) = \frac{6}{36} = \frac{1}{6}$.
    e. $P(F|E) = \dfrac{P(E \cap F)}{P(E)} = \dfrac{\frac{1}{36}}{\frac{1}{6}} = \dfrac{1}{6} \neq P(F) = \dfrac{1}{12}$ and so the events are not independent.

17. Let A denote the event that the sum of the numbers is less than 9 and let B denote the
    event that one of the numbers is a 6. Then, $P(A|B) = \dfrac{P(A \cap B)}{P(B)} = \dfrac{\frac{4}{36}}{\frac{11}{36}} = \dfrac{4}{11}$.

19. Refer to Figure 7.15, page 416 in the text. Here
$$E = \{(3,1),(3,2),(3,3),(3,4),(3,5),(3,6)\}$$
and $F = \{(1,6),(6,1),(2,5),(5,2),(3,4),(4,3)\}$. Then $E \cap F = \{(3,4)\}$.

Now, $P(E \cap F) = \dfrac{1}{36}$ and this is equal to $P(E) \cdot P(F) = \left(\dfrac{6}{36}\right)\left(\dfrac{6}{36}\right) = \dfrac{1}{36}$.

So E and F are independent events.

21. $P(E \cap F) = \frac{13}{24} = \frac{1}{4}$, $P(E) = \frac{26}{52} = \frac{1}{2}$, and $P(F) = \frac{13}{52} = \frac{1}{4}$. Now,
$$P(E) \cdot P(F) = (\tfrac{1}{2})(\tfrac{1}{4}) = \tfrac{1}{8} \neq P(E \cap F) = \tfrac{1}{4}.$$

So E and F are not independent events. The knowledge that the card drawn is black increases the probability that it is a spade.

23. Let A denote the event that the battery lasts 10 or more hours and let B denote the event that the battery lasts 15 or more hours.
Then $P(A) = 0.8, \quad P(B) = 0.15$
and $P(A \cap B) = 0.15$.

Therefore, the probability that the battery will last 15 hours or more is
$$P(B|A) = \frac{P(A \cap B)}{P(A)} = \frac{0.15}{0.8} = \frac{3}{16} = 0.1875.$$

25. Refer to the following tree diagram:

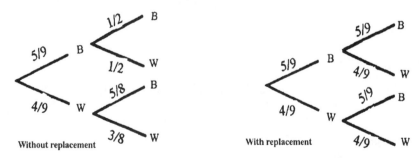

Without replacement   With replacement

a. The probability that the second ball drawn is a white ball if the second ball is drawn without replacing the first is
$$P(B)P(W|B) + P(W)P(W|W) = (\tfrac{5}{9})(\tfrac{1}{2}) + (\tfrac{4}{9})(\tfrac{3}{8}) = \tfrac{4}{9}.$$

b. The probability that the second ball drawn is a white ball if the first ball is replaced before the second is drawn is $(\tfrac{5}{9})(\tfrac{4}{9}) + (\tfrac{4}{9})(\tfrac{4}{9}) = \tfrac{4}{9}.$

27. Refer to the following tree diagram:

a. The probability that a student selected at random from this medical school is black is $(\frac{1}{7})(\frac{1}{3}) = \frac{1}{21}$

b. The probability that a student selected at random from this medical school is black if it is known that the student is a member of a minority group is $P(B|M) = 1/3$.

29. Let D denote the event that the card drawn is a diamond. Consider the tree diagram that follows.

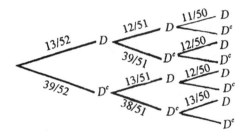

Then the required probability is
$$(\frac{13}{52})(\frac{12}{51})(\frac{11}{50}) + (\frac{13}{52})(\frac{39}{51})(\frac{12}{50}) + (\frac{39}{52})(\frac{13}{51})(\frac{12}{50}) + (\frac{39}{52})(\frac{38}{51})(\frac{13}{50}) = 0.25.$$

31. The sample space for a three-child family is
   $S = \{GGG, GGB, GBG, GBB, BGG, BGB, BBG, BBB\}$.
   Since we know that there is at least one girl in the three-child family we are dealing with a reduced sample space
   $S_1 = \{GGG, GGB, GBG, GBB, BGG, BGB, BBG\}$
   in which there are 7 outcomes. Then the probability that all three children are girls is
   $$P(E) = \frac{n(E)}{n(S)} = \frac{1}{7}.$$

33. a.

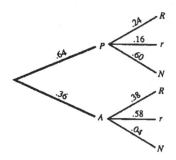

P = Professional
A = Amateur
R = recovered within 48 hrs
r = recovered after 48 hrs
N = never recovered

b. The required probability is 0.24.
c. The required probability is $(0.64)(0.60) + (0.36)(0.04) \approx 0.40$.

35. Refer to the following tree diagram.

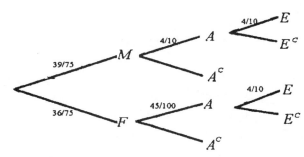

a. $P(A \cap E \mid M) = (0.4)(0.4) = 0.16$.

b. $P(A) = P(M \cap A) + P(F \cap A) = \dfrac{39}{75} \cdot \dfrac{4}{10} + \dfrac{36}{75} \cdot \dfrac{45}{100} = 0.424$

c. $P(M \cap A \cap E) + P(F \cap A \cap E) = \dfrac{39}{75} \cdot \dfrac{4}{10} \cdot \dfrac{4}{10} + \dfrac{36}{75} \cdot \dfrac{45}{100} \cdot \dfrac{4}{10}$

$$= 0.0832 + 0.0864 = 0.1696.$$

37. a. The probability that none of the dozen eggs is broken is $(0.992)^{12} = 0.908$.
Therefore, the probability that at least one egg is broken is $1 - 0.908 = 0.092$.
b. Using the results of (a), we see that the required probability is
$$(0.092)(0.092)(0.908) \approx 0.008.$$

39. a. $P(A) = \dfrac{1120}{4000} = 0.280$; $P(B) = \dfrac{1560}{4000} = 0.390$;

$P(A \cap B) = \dfrac{720}{4000} = 0.180$;

$P(B|A) = \dfrac{P(A \cap B)}{P(A)} = \dfrac{n(A \cap B)}{n(A)} = \dfrac{720}{1120} \approx 0.643$

$P(B|A^C) = \dfrac{P(A^C \cap B)}{P(A^C)} = \dfrac{n(A^C \cap B)}{n(A^C)} = \dfrac{840}{2880} \approx 0.292$.

b. $P(B|A) \neq P(B)$ so A and B are not independent events.

41. Let C denote the event that a person in the survey was a heavy coffee drinker and Pa denote the event that a person in the survey had cancer of the pancreas. Then
$$P(C) = \dfrac{3200}{10000} = 0.32 \text{ and } P(Pa) = \dfrac{160}{10000} = 0.016$$

$$P(C \cap Pa) = \dfrac{132}{160} = 0.825$$
and $P(C) \cdot P(Pa) = 0.00512 \neq P(C \cap Pa)$. Therefore the events are not independent.

43. The probability that the first test will fail is 0.03, that the second test will fail is 0.015, and the third test will fail is 0.015. Since these are independent events the probability that all three tests will fail is
$$(0.03)(0.015)(0.015) = 0.0000068.$$

45. a. Let $P(A)$, $P(B)$, $P(C)$ denote the probability that the first, second, and third patient suffer a rejection, respectively. Then $P(A) = \dfrac{1}{2}$, $P(B) = \dfrac{1}{3}$, and $P(C) = \dfrac{1}{10}$.
Therefore, the probabilities that each patient does not suffer a rejection are given by
$P(A^c) = \dfrac{1}{2}$, $P(B^c) = \dfrac{2}{3}$, and $P(C^c) = \dfrac{9}{10}$. Then, the probability that none of the 3 patients suffers a rejection is given by

$$P(A^c) \cdot P(B^c) \cdot P(C^c) = \frac{1}{2} \cdot \frac{2}{3} \cdot \frac{9}{10} = \frac{18}{60} = \frac{3}{10}.$$

Therefore, the probability that at least one patient will suffer rejection is

$$1 - P(A^c) \cdot P(B^c) \cdot P(C^c) = 1 - \frac{3}{10} = \frac{7}{10}.$$

b. The probability that exactly two patients will suffer rejection is

$$P(A)P(B)P(C^c) + P(A)P(B^c)P(C) + P(A^c)P(B)P(C)$$

$$= \frac{1}{2} \cdot \frac{1}{3} \cdot \frac{9}{10} + \frac{1}{2} \cdot \frac{2}{3} \cdot \frac{1}{10} + \frac{1}{2} \cdot \frac{1}{3} \cdot \frac{1}{10} = \frac{9+2+1}{60} = \frac{12}{60} = \frac{1}{5}.$$

47. Let A denote the event that at least one of the floodlights remain functional over the one-year period. Then

$$P(A) = 0.99999 \text{ and } P(A^c) = 1 - P(A) = 0.00001.$$

Letting n represent the minimum number of floodlights needed, we have

$$(0.01)^n = 0.00001$$

$$n \log(0.01) = -5$$

$$n(-2) = -5$$

$$n = \frac{5}{2} = 2.5.$$

Therefore, the minimum number of floodlights needed is 3.

49. The probability that the event will not occur in one trial is $1 - p$. Therefore, the probability that it will not occur in n independent trials is $(1 - p)^n$. Therefore, the probability that it will occur at least once in n independent trials is $1 - (1 - p)^n$.

51. $P(E|F) = \dfrac{P(E \cap F)}{P(F)} = \dfrac{P(F)}{P(F)} = 1 \qquad (E \cap F = F \text{ since } F \subset E).$

Interpretation: Since $F \subset E$, an occurrence of F implies an occurrence of E. In other words, given that F has occurred, it is a certainty that E will occur, that is, $P(E|F) = 1$.

53. True. Since A and B are mutually exclusive, $A \cap B = \varnothing$, and

$$P(A|B) = \frac{P(A \cap B)}{P(B)} = \frac{P(\varnothing)}{P(B)} = 0.$$

55. True. This follows from Formula 3, $P(A \cap B) = P(A) \cdot P(B|A).$

1.

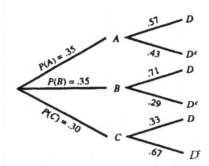

3.   a. $P(D^C) = \dfrac{15+10+20}{35+35+30} = 0.45$;   b. $P(B|D^C) = \dfrac{10}{15+10+20} = 0.22$.

5.   a. $P(D) = \dfrac{25+20+15}{50+40+35} = 0.48$ b. $P(B|D) = \dfrac{20}{25+20+15} = 0.33$

7.   a. $P(A) \cdot P(D|A) = (0.4)(0.2) = 0.08$ b. $P(B) \cdot P(D|B) = (0.6)(0.25) = 0.15$

   c. $P(A|D) = \dfrac{P(A) \cdot P(D|A)}{P(A) \cdot P(D|A) + P(B) \cdot P(D|B)} = \dfrac{(0.4)(0.2)}{0.08 + 0.15} \approx 0.35$

9.   a. $P(A) \cdot P(D|A) = \dfrac{1}{3} \cdot \dfrac{1}{4} = \dfrac{1}{12}$ b. $P(B) \cdot P(D|B) = \dfrac{1}{2} \cdot \dfrac{1}{2} = \dfrac{1}{4}$

   c. $P(C) \cdot P(D|C) = \dfrac{1}{6} \cdot \dfrac{1}{3} = \dfrac{1}{18}$

   d. $P(A|D) = \dfrac{P(A) \cdot P(D|A)}{P(A) \cdot P(D|A) + P(B) \cdot P(D|B) + P(C) \cdot P(C|B)}$

   $= \dfrac{\frac{1}{12}}{\frac{1}{12} + \frac{1}{4} + \frac{1}{18}} = \dfrac{1}{12} \cdot \dfrac{36}{14} = \dfrac{3}{14}$

11. Let A denote the event that the first card drawn is a heart and B the event that the second card drawn is a heart. Then

$$P(A|B) = \frac{P(A) \cdot P(B|A)}{P(A) \cdot P(B|A) + P(A^C) \cdot P(B|A^C)}$$

$$= \frac{\frac{1}{4} \cdot \frac{12}{51}}{\frac{1}{4} \cdot \frac{12}{51} + \frac{3}{4} \cdot \frac{13}{51}} = \frac{4}{17}.$$

13. Using the following tree diagram, we see that

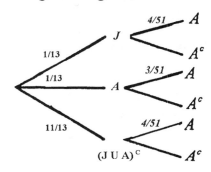

$$P(J|A) = \frac{\frac{1}{13} \cdot \frac{4}{51}}{\frac{1}{13} \cdot \frac{4}{51} + \frac{1}{13} \cdot \frac{3}{51} + \frac{11}{13} \cdot \frac{4}{51}} = \frac{\frac{4}{13 \cdot 51}}{\frac{51}{13 \cdot 51}} = 0.0784 \,.$$

15. The probabilities associated with this experiment are represented in the following tree diagram.

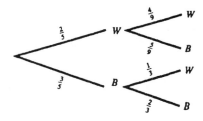

17. Referring to the tree diagram in Exercise 15, we see that the probability that the transferred ball was black given that the second ball was white is

$$P(B|W) = \frac{\frac{3}{5} \cdot \frac{1}{3}}{\frac{2}{5} \cdot \frac{4}{9} + \frac{3}{5} \cdot \frac{1}{3}} = \frac{9}{17} \,.$$

19. Let D denote the event that a senator selected at random is a Democrat, R denote the event that a senator selected at random is a Republican, and M the event that a

senator has served in the military. From the following tree diagram

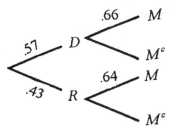

we see that the probability that a senator selected at random who has served in the
military is a Republican is

$$P(R|M) = \frac{P(R)P(M|R)}{P(M)} = \frac{(0.64)(0.43)}{(0.66)(0.57)+(0.64)(0.43)}$$

$$\approx 0.422.$$

21. Let H_2 denote the event that the coin tossed is the two-headed coin, H_B denote the
event that the coin tossed is the biased coin, and H_F denote the event that the coin
tossed is the fair coin. Referring to the following tree diagram, we see that

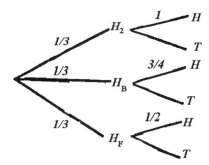

a. $P(H) = \dfrac{1}{3} \cdot 1 + \dfrac{1}{3} \cdot \dfrac{3}{4} + \dfrac{1}{3} \cdot \dfrac{1}{2} = \dfrac{1}{3} + \dfrac{1}{4} + \dfrac{1}{6} = \dfrac{9}{12} = \dfrac{3}{4}$

b. $P(H_F|H) = \dfrac{\frac{1}{3} \cdot \frac{1}{2}}{\frac{3}{4}} = \dfrac{2}{9}$

23. Let D denote the event that the person has the disease, and let Y denote the event that
the test is positive. Referring to the following tree diagram, we see that the required
probability is

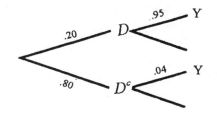

$$P(D|Y) = \frac{P(D) \cdot P(Y|D)}{P(D) \cdot P(Y|D) + P(D^C) \cdot P(Y|D^C)}$$

$$= \frac{(0.2)(0.95)}{(0.2)(0.95) + (0.8)(0.04)} \approx 0.856.$$

25. Let x denote the age of an insured driver, and let A denote the event that an insured driver is in an accident. Using the tree diagram we find,

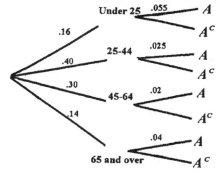

   a. $P(A) = (0.16)(0.055) + (0.4)(0.025) + (0.3)(0.02) + (0.14)(0.04)$
         $\approx 0.03.$

   b. $P(x < 25 | A) = \dfrac{(0.16)(0.055)}{(0.03)} \approx 0.29$

27. Let E, F, and G denote the events that the child selected at random is 12 years old, 13 years old, or 14 years old, respectively; and let C^c denote the event that the child does not have a cavity. Using the tree diagram that follows, we see that

$$P(G|C^c) = \frac{\frac{5}{12}(0.28)}{\frac{1}{4}(0.42) + \frac{1}{3}(0.34) + \frac{5}{12}(0.28)} = 0.348$$

                    7   *Probability*

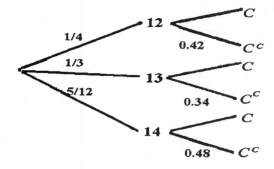

29. Let M and F denote the events that a person arrested for crime in 1988 was male or female, respectively; and let U denote the event that the person was under the age of 18. Using the following tree diagram, we have

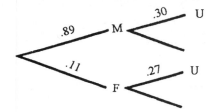

a. $P(U) = (0.89)(0.30) + (0.11)(0.27) = 0.2967$.

b. $P(F|U) = \dfrac{(0.11)(0.27)}{(0.89)(0.30)+(0.11)(0.27)} = 0.1001$.

31. Let I and II denote a customer who purchased the drug in capsule or tablet form, respectively; and let E denote a customer in Group I and II who purchased the extra-strength dosage of the drug. Then using the tree diagram that follows, we see that

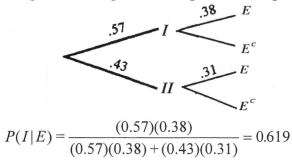

$$P(I|E) = \frac{(0.57)(0.38)}{(0.57)(0.38)+(0.43)(0.31)} = 0.619$$

33. Let N and D denote the events that a employee was placed by Nancy or Darla, respectively; and let S denote the event that the employee placed by one of these women was satisfactory. Using the tree diagram that follows, we see that

$$P(D|S^C)=\frac{(0.55)(0.3)}{(0.45)(0.2)+(0.55)(0.3)} = 0.647$$

```
                      .80      S
              N  <
    .45              .20
  <                           S^c
    .55              .70      S
              D  <
                     .30
                              S^c
```

35. Using the tree diagram shown at the right, we see that

$$P(S_2|S) = \frac{(0.95)(0.8)}{(0.95)(0.8)+(0.2)(0.3)} \approx 0.93.$$

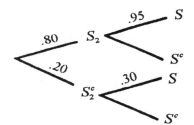

37. Let A, B, C, and D denote the event that the age of a guest is between 21 and 34, between 35 and 44, ..., 55 and over, respectively, and let O denote the event that a man keeps his paper money in order of denomination. Refer to the tree diagram below.

```
        25 = 5       .9      O
        80   16   A <
                            O^c
        30 = 3      .61     O
        80   8    B <
                            O^c
        15 = 3      .8      O
        80   16   C <
                            O^c
        10 = 1      .8      O
        80   8    D <
                            O^c
```

The required probability is

$$P(B|O) = \frac{\left(\frac{3}{8}\right)\left(\frac{61}{100}\right)}{\left(\frac{5}{16}\right)\left(\frac{9}{10}\right)+\left(\frac{3}{8}\right)\left(\frac{61}{100}\right)+\left(\frac{3}{16}\right)\left(\frac{8}{10}\right)+\left(\frac{1}{8}\right)\left(\frac{8}{10}\right)} \approx 0.3010.$$

39. Let D and R denote the event that the respondent is a Democrat or a Republican voter, respectively. Next, let S, O, and R denote the event that the respondent supports, opposes, or either doesn't know or refuses, respectively. Refer to the following diagram

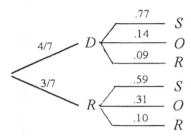

The required probability is

$$P(O|D) = \frac{\left(\frac{4}{7}\right)(0.14)}{\left(\frac{4}{7}\right)(0.14)+\left(\frac{3}{7}\right)(0.31)} = \frac{(0.571428)(0.14)}{(0.571428)(0.14)+(0.428571)(0.31)}$$
$$\approx 0.3758.$$

41. Let A, B, C, D, and E denote the event that the income level is 0-16, 17-33, ..., 96-100 percentile, respectively, and let V denote the event that a person voted in the election. Refer to the following diagram.

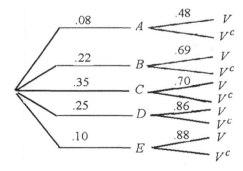

We see that the probability is

$$P(V|B) = \frac{(0.22)(0.69)}{(0.08)(0.48)+(0.22)(0.60)+(0.35)(0.70)+(0.25)(0.86)+(0.10)(0.88)}$$
$$\approx 0.2056.$$

1. a. $P(E \cap F) = 0$ since E and F are mutually exclusive.
   b. $P(E \cup F) = P(E) + P(F) - P(E \cap F) = 0.4 + 0.2 = 0.6$.
   c. $P(E^c) = 1 - P(E) = 1 - 0.4 = 0.6$.
   d. $P(E^C \cap F^C) = P(E \cup F)^C = 1 - P(E \cup F) = 1 - 0.6 = 0.4$.
   e. $P(E^C \cup F^C) = P(E \cap F)^C = 1 - P(E \cap F) = 1 - 0 = 1$.

3. a. The probability of the number being even is
      $P(2) + P(4) + P(6) = 0.12 + 0.18 + 0.19 = 0.49$.
   b. The probability that the number is either a 1 or a 6 is
      $P(1) + P(6) = 0.20 + 0.19 = 0.39$.
   c. The probability that the number is less than 4 is
      $P(1) + P(2) + P(3) = 0.20 + 0.12 + 0.16 = 0.48$.

5. Let A denote the event that a video-game cartridge has an audio defect and let V denote the event that a video-game cartridge has a video defect. Then using the following Venn diagram, we have

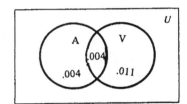

   a. the probability that a cartridge purchased by a customer will have a video or audio defect is
   $$P(A \cup V) = P(A \cap V^C) + P(V \cap A^C) + P(A \cap V)$$
   $$= 0.004 + 0.004 + 0.011 = 0.019$$
   b. the probability that a cartridge purchased by a customer will have not have a video or audio defect is
   $$P(A \cup V)^C = 1 - P(A \cup V) = 1 - 0.019 = 0.981.$$

7. $P(A \cap E) = (0.3)(0.6) = 0.18$.   9. $P(C \cap E) = (0.2)(0.3) = 0.06$.

11. $P(E) = 0.18 + 0.25 + 0.06 = 0.49$

13. a. The probability that none of the pens in the sample are defective is
$$\frac{C(18,3)}{C(20,3)} = \frac{\dfrac{18!}{15!3!}}{\dfrac{20!}{17!3!}} = \frac{18!}{15!}\cdot\frac{17!}{20!} = \frac{68}{95} \approx 0.71579.$$
Therefore, the probability that at least one is defective is given by
$$1 - 0.71579 \approx 0.284.$$
b. The probability that two are defective is given by
$$\frac{C(2,2)\cdot C(18,1)}{C(20,3)} = \frac{18}{\dfrac{20!}{17!3!}} = \frac{6}{380} \approx 0.0158 \ .$$
Therefore, the probability that no more than 1 is defective is given by
$$1 - \frac{6}{380} = \frac{374}{380} \approx 0.984 \ .$$

15. Let E denote the event that the sum of the numbers is 8 and let D denote the event that the numbers appearing on the face of the two dice are different. Then
$$P(E|D) = \frac{P(E\cap D)}{P(D)} = \frac{4}{30} = \frac{2}{15}.$$

17. Let E, F, and G denote the events that the first toss results in an even number being shown, the second toss results in an odd number being shown, and the third toss results in a one being shown, respectively. Then $P(E) = \frac{1}{6}$, $P(F) = \frac{1}{2}$, and $P(G) = \frac{12}{52} = \frac{3}{13}$. Since the outcomes are independent, the required probability is
$$P(E\cap F\cap G) = P(E)P(F)P(G) = (\tfrac{1}{6})(\tfrac{1}{2})(\tfrac{3}{13}) = \tfrac{1}{52}.$$

19. The probability that all three cards are face cards is $\dfrac{C(12,3)}{C(52,3)} = 0.00995.$

21. Referring to the tree diagram at the right, we see that the probability that the second card is black, given that the first card was red is
$$\frac{26}{51} = 0.510 \ .$$

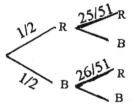

23. Let M denote the event that an employee at the insurance company is a male and let F denote the event that an employee at the insurance company is on flex time.Then

$$P(M|F) = \frac{\frac{320}{600} \cdot \frac{160}{320}}{\frac{320}{600} \cdot \frac{160}{320} + \frac{280}{600} \cdot \frac{190}{280}} = 0.4571.$$

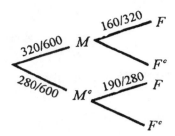

25. Let E denote the event that an applicant selected at random is eligible for admission, and let Pa denote the event that an applicant selected at random passes the admission exam. Using the tree diagram that follows, we see that

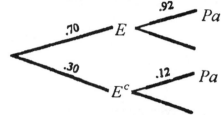

a. $P(Pa) = (0.70)(0.92) + (0.30)(0.12) = 0.68.$

b. $P(E^C|Pa) = \dfrac{(0.30)(0.12)}{(0.70)(0.92)+(0.30)(0.12)} = 0.053.$

7   *Probability*

CHAPTER 8

EXERCISES 8.1, page 446

1. a. See part (b).
    b.   c. $\{GGG\}$

Outcome	GGG	GGR	GRG	RGG	GRR	RGR	RRG	RRR
Value	3	2	2	2	1	1	1	0

3. X may assume the values in the set $S = \{1,2,3,...\}$.

5. The event that the sum of the dice is 7 is $E = \{(1,6),(2,5),(3,4),(4,3),(5,2),(6,1)\}$
    and $P(E) = \dfrac{6}{36} = \dfrac{1}{6}$.

7. X may assume the value of any positive integer. The random variable is infinite
    discrete.

9. $\{d\,|\,d \geq 0\}$. The random variable is continuous.

11. X may assume the value of any positive integer. The random variable is infinite
    discrete.

13. a. $P(X = -10) = 0.20$ b. $P(X \geq 5) = 0.1 + 0.25 + 0.1 + 0.15 = 0.60$
    c. $P(-5 \leq X \leq 5) = 0.15 + 0.05 + 0.1 = 0.30$
    d. $P(X \leq 20) = 0.20 + 0.15 + 0.05 + 0.1 + 0.25 + 0.1 + 0.15 = 1$

15.

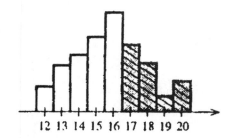

17. a.

x	1	2	3	4	5	6
$P(X=x)$	$\frac{1}{6}$	$\frac{1}{6}$	$\frac{1}{6}$	$\frac{1}{6}$	$\frac{1}{6}$	$\frac{1}{6}$

y	1	2	3	4	5	6
$P(Y=y)$	$\frac{1}{6}$	$\frac{1}{6}$	$\frac{1}{6}$	$\frac{1}{6}$	$\frac{1}{6}$	$\frac{1}{6}$

b.

$x+y$	2	3	4	5	6	7	8	9	10	11	12
$P(X+Y)=x+y$	$\frac{1}{36}$	$\frac{2}{36}$	$\frac{3}{36}$	$\frac{4}{36}$	$\frac{5}{36}$	$\frac{6}{36}$	$\frac{5}{36}$	$\frac{4}{36}$	$\frac{3}{36}$	$\frac{2}{36}$	$\frac{1}{36}$

19. a.

x	0	1	2	3	4	5
$P(X=x)$	0.017	0.067	0.033	0.117	0.233	0.133

x	6	7	8	9	10
$P(X=x)$	0.167	0.1	0.05	0.067	0.017

b.

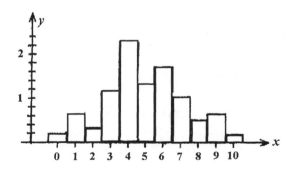

21.

x	1	2	3	4	5
$P(X=x)$	0.007	0.029	0.021	0.079	0.164

x	6	7	8	9	10
P(X=x)	0.15	0.20	0.207	0.114	0.029

23. True. This follows from the definition.

USING TECHNOLOGY EXERCISES 8.1, page 450

1. Data from Table 8.1, page 442 in the text   3.

EXERCISES 8.2, page 460

1. a. The student's grade-point average is given by

$$\frac{(2)(4)(3) + (3)(3)(3) + (4)(2)(3) + (1)(1)(3)}{(10)(3)} \text{ or } 2.6.$$

   b.

X	0	1	2	3	4
P(X = x)	0	0.1	0.4	0.3	0.2

   $E(X) = 1(0.1) + 2(0.4) + 3(0.3) + 4(0.2) = 2.6.$

3. $E(X) = -5(0.12) + -1(0.16) + 0(0.28) + 1(0.22) + 5(0.12) + 8(0.1)$
   $= 0.86.$

5. $E(X) = 0(0.07) + 25(0.12) + 50(0.17) + 75(0.14) + 100(0.28)$
   $+ 125(0.18) + 150(0.04)$
   $= 78.5, \text{ or } \$78.50.$

7. A customer entering the store is expected to buy

$E(X) = (0)(0.42) + (1)(0.36) + (2)(0.14) + (3)(0.05) + (4)(0.03)$
$= 0.91$, or 0.91 videocassettes.

9.   The expected number of accidents is given by
$E(X) = (0)(0.935) + (1)(0.03) + (2)(0.02) + (3)(0.01) + (4)(0.005) = 0.12$

11. The expected number of machines that will break down on a given day is given by
$E(X) = (0)(0.43) + (1)(0.19) + (2)(0.12) + (3)(0.09) + (4)(0.04)$
$\quad\quad + (5)(0.03) + (6)(0.03) + (7)(0.02) + (8)(0.05)$
$= 1.73$.

13. The associated probabilities are $\frac{3}{50}, \frac{8}{50}, ..., \frac{5}{50}$, respectively. Therefore, the expected interest rate is
$(4.9)(\frac{3}{50}) + 5(\frac{8}{50}) + 5.1(\frac{12}{50}) + 5.2(\frac{14}{50}) + 5.3(\frac{8}{50}) + 5.4(\frac{5}{50})$
≈ 5.162,   or 5.16%

15. The expected net earnings of a person who buys one ticket are
$-1(0.997) + 24(0.002) + 99(0.0006) + 499(0.0002) + 1999(0.0002)$
$= -0.39$, or a loss of $0.39 per ticket.

17. The expected gain of the insurance company is given by
$E(X) = 0.992(130) - (9870)(0.008) = 50$, or $50.

19. The company's expected gross profit is given by
$E(X) = (0.3)(80,000) + (0.60)(75,000) + (0.10)(70,000) - 64,000$
$= 12,000$, or $12,000.

21. City A: $E(X) = (10,000,000)(0.2) - 250,000 = 1,750,000$,   or $1.75 million.
City B: $E(X) = (7,000,000)(0.3) - 200,000 = 1,900,000$, or $1.9 million.
We see that the company should bid for the rights in city B.

23. The expected number of houses sold per year at company A is given by
$E(X) = (12)(.02) + (13)(.03) + (14)(.05) + (15)(.07) + (16)(.07)$
$\quad\quad + (17)(.16) + (18)(.17) + (19)(.13) + (20)(.11)$
$\quad\quad + (21)(.09) + (22)(.06) + (23)(.03) + (24)(.01)$
$= 18.09$.

The expected number of houses sold per year at company B is given by

$$E(X) = (6)(.01) + (7)(.04) + (8)(.07) + (9)(.06) + (10)(.11)$$
$$+ (11)(.12) + (12)(.19) + (13)(.17) + (14)(.13)$$
$$+ (15)(.04) + (16)(.03) + (17)(.02) + (18)(.01)$$
$$= 11.77.$$

Then, Sally's expected commission at company A is given by
$$(0.03)(154{,}000)(18.09) = 83{,}575.80, \text{ or } \$83{,}575.80$$
Her expected commission at company B is given by
$$(0.03)(237{,}000)(11.77) = 83{,}684.70,$$
or $83,684.70. Based on these expectations, she should accept the job offer with company B.

25. Maria might expect her business to grow at the rate of
$$(5)(0.12) + (4.5)(0.24) + (3)(0.4) + (0)(0.2) + (-0.5)(0.04) = 2.86$$
or 2.86%/year for the upcoming year.

27. The expected value of the winnings on a $1 bet placed on a split is
$$E(X) = 17 \cdot \frac{2}{38} + (-1) \cdot \frac{36}{38} \approx -0.0526, \text{ or a loss of 5.3 cents.}$$

29. The expected value of a player's winnings are
$$(1)(\frac{18}{37}) + (-1)(\frac{19}{37}) = -\frac{1}{37} \approx -0.027, \text{ or a loss of 2.7 cents per bet.}$$

31. The odds in favor of E occurring are $\dfrac{P(E)}{P(E^C)} = \dfrac{0.4}{0.6}$, or 2 to 3. The odds against E occurring are 3 to 2.

33. The probability of E not occurring is given by $P(E) = \dfrac{2}{3+2} = \dfrac{2}{5} = 0.4.$

35. The probability that she will win her match is $P(E) = \dfrac{7}{7+5} = \dfrac{7}{12} = 0.5833.$

37. The probability that the business deal will not go through is
$$P(E) = \frac{5}{5+9} = \frac{5}{14} \approx 0.3571.$$

39. a. The mean is given by

$$\frac{40 + 45 + 2(50) + 55 + 2(60) + 2(75) + 2(80) + 4(85) + 2(90) + 2(95) + 100}{20}$$

   = 74.

   The mode is 85 (the value that appears most frequently).
   The median is 80 (the middle value).
   b. The mode is the least representative of this set of test scores.

41. True This follows from the definition.

EXERCISES 8.3, page 471

1. $\mu = (1)(0.4) + (2)(0.3) + 3(0.2) + (4)(0.1) = 2.$
    $\text{Var}(X) = (0.4)(1 - 2)^2 + (0.3)(2 - 2)^2 + (0.2)(3 - 2)^2 + (0.1)(4 - 2)^2$
    $\qquad = 0.4 + 0 + 0.2 + 0.4 = 1$
    $\sigma = \sqrt{1} = 1.$

3. $\mu = -2(\frac{1}{16}) + -1(\frac{4}{16}) + 0(\frac{6}{16}) + 1(\frac{4}{16}) + 2(\frac{1}{16}) = \frac{0}{16} = 0.$
    $\text{Var}(X) = \frac{1}{16}(-2-0)^2 + \frac{4}{16}(-1-0)^2 + \frac{6}{16}(0-0)^2 + \frac{4}{16}(1-0)^2 + \frac{1}{16}(2-0)^2$
    $\qquad = 1$
    $\sigma = \sqrt{1} = 1.$

5. $\mu = 0.1(430) + (0.2)(480) + (0.4)(520) + (0.2)(565) + (0.1)(580)$
    $\qquad = 518.$
    $\text{Var}(X) = 0.1(430 - 518)^2 + (0.2)(480 - 518)^2 + (0.4)(520 - 518)^2$
    $\qquad\qquad + (0.2)(565 - 518)^2 + (0.1)(580 - 518)^2$
    $\qquad = 1891.$
    $\sigma = \sqrt{1891} \approx 43.49.$

7. The mean of the histogram in Figure (b) is more concentrated about its mean than
    the histogram in Figure (a). Therefore, the histogram in Figure (a) has the larger
    variance.

9. $E(X) = 1(0.1) + 2(0.2) + 3(0.3) + 4(0.2) + 5(0.2) = 3.2.$

$$\text{Var}(X) = (0.1)(1 - 3.2)^2 + (0.2)(2 - 3.2)^2 + (0.3)(3 - 3.2)^2$$
$$+ (0.2)(4 - 3.2)^2 + (0.2)(5 - 3.2)^2$$
$$= 1.56$$

11. $\mu = \dfrac{1 + 2 + 3 + \cdots + 8}{8} = 4.5$

$V(X) = \frac{1}{8}(1 - 4.5)^2 + \frac{1}{8}(2 - 4.5)^2 + \cdots + \frac{1}{8}(8 - 4.5)^2 = 5.25$

13. a. Let X be the annual birth rate during the years 1981 - 1990.

b.

x	15.5	15.6	15.7	15.9	16.2	16.7
$P(X = x)$	0.2	0.1	0.3	0.2	0.1	0.1

c. $E(X) = (0.2)(15.5) + (0.1)(15.6) + (0.3)(15.7) + (0.2)(15.9)$
$$+ (0.1)(16.2) + (0.1)(16.7)$$
$$= 15.84.$$
$V(X) = (0.2)(15.5 - 15.84)^2 + (0.1)(15.6 - 15.84)^2$
$$+ (0.3)(15.7 - 15.84)^2 + (0.2)(15.9 - 15.84)^2$$
$$+ (0.1)(16.2 - 15.84)^2 + (0.1)(16.7 - 15.84)^2$$
$$= 0.1224$$
$$\sigma = \sqrt{0.1224} \approx 0.350.$$

15. a. Mutual Fund A

$\mu = (0.2)(-4) + (0.5)(8) + (0.3)(10) = 6.2$, or $620.
$V(X) = (0.2)(-4 - 6.2)^2 + (0.5)(8 - 6.2)^2 + (0.3)(10 - 6.2)^2$
$$= 26.76, \text{ or } \$267{,}600.$$

Mutual Fund B

$\mu = (0.2)(-2) + (0.4)(6) + (0.4)(8)$
$$= 5.2, \text{ or } \$520.$$
$V(X) = (0.2)(-2 - 5.2)^2 + (0.4)(6 - 5.2)^2 + (0.4)(8 - 5.2)^2$
$$= 13.76, \text{ or } \$137{,}600.$$

b. Mutual Fund A c. Mutual Fund B

17. $\text{Var}(X) = (0.4)(1)^2 + (0.3)(2)^2 + (0.2)(3)^2 + (0.1)(4)^2 - (2)^2 = 1.$

19. $\mu = [\frac{10}{500}(180) + \frac{20}{500}(190) + \cdots + \frac{5}{500}(350)] = 239.6$, or $239,600.

$$V(X) = [\tfrac{10}{500}(180-239.6)^2 + \tfrac{20}{500}(190-239.66)^2 + \cdots + \tfrac{5}{500}(350-239.66)^2][(1000)^2]$$
$$=1443.84 \times 10^6 \text{ dollars.}$$

$$\sigma = \sqrt{1443.84 \times 10^6} = 37.998 \times 10^3, \text{ or } \$37,998.$$

21. The mean is given by
$$22\left(\tfrac{1332}{27127}\right)+27\left(\tfrac{4219}{27127}\right)+32\left(\tfrac{6345}{27127}\right)+37\left(\tfrac{7598}{27127}\right)+42\left(\tfrac{7633}{27127}\right) \approx 34.9456, \text{ or } 34.95.$$
$$\text{Var}(x) = \left(\tfrac{1332}{27127}\right)(22-34.9456)^2 + \left(\tfrac{4219}{27127}\right)(27-34.9456)^2 + \cdots$$
$$+\left(\tfrac{7633}{27127}\right)(42-34.9456)^2 \approx 35.26222.$$
Therefore, $\sigma = 5.938$, or 5.94.

23. a. Using Chebychev's inequality we have
$$P(\mu - k\sigma \le X \le \mu + k\sigma) \ge 1 - 1/k^2.$$
$$\mu - k\sigma = 42 - k(2) = 38, \text{ and } k = 2,$$
and $\quad P(\mu - k\sigma \le X \le \mu + k\sigma) \ge 1 - 1/(2)^2$
$$\ge 1 - 1/4$$
$$\ge 3/4, \text{ or at least } 0.75.$$
b. Using Chebychev's inequality we have
$$P(\mu - k\sigma \le X \le \mu + k\sigma) \ge 1 - 1/k^2.$$
$$\mu - k\sigma = 42 - k(2) = 32, \text{ and } k = 5,$$
and $\quad P(\mu - k\sigma \le X \le \mu + k\sigma) \ge 1 - 1/(5)^2$
$$\ge 1 - 1/25 \ge 24/25, \text{ or at least } 0.96.$$

25. Here $\mu = 50$ and $\sigma = 1.4$. Now, we require that $c = k\sigma$, or $k = \dfrac{c}{1.4}$.

Next, we solve $\quad 0.96 = 1 - \left(\dfrac{1.4}{c}\right)^2$; $\quad \dfrac{1.96}{c^2} = 0.04$; $\quad c^2 = \dfrac{1.96}{0.04} = 49$, or $c = 7$.

27. Using Chebychev's inequality we have $P(\mu - k\sigma \le X \le \mu + k\sigma) \ge 1 - 1/k^2$.
Here, $\mu - k\sigma = 24 - k(3) = 20$, and $k = 4/3$.
So $\quad P(\mu - k\sigma \le X \le \mu + k\sigma) \ge 1 - 1/(4/3)^2 \ge 1 - 9/16 = 7/16$, or at least 0.4375.

29. Using Chebychev's inequality we have
$$P(\mu - k\sigma \le X \le \mu + k\sigma) \ge 1 - 1/k^2.$$
Here, $\mu - k\sigma = 42,000 - k(500) = 40,000$, and $k = 4$.

So $\quad P(\mu - k\sigma \le X \le \mu + k\sigma) \ge 1 - 1/(4)^2 \ge 1 - 1/16 \ge 15/16,$ or at least 0.9375.

31. True. This follows from the definition.

USING TECHNOLOGY EXERCISES, page 475

1. a.

3. a.

b. $\mu = 4$ and $\sigma \approx 1.40$

b. $\mu = 17.34$ and $\sigma \approx 1.11$

5. a. Let X denote the random variable that gives the weight of a carton of sugar.
    b. The probability distribution for the random variable X is

x	4.96	4.97	4.98	4.99	5.00	5.01	5.02	5.03	5.04	5.05	5.06
$P(X = x)$	$\frac{3}{30}$	$\frac{4}{30}$	$\frac{4}{30}$	$\frac{1}{30}$	$\frac{1}{30}$	$\frac{5}{30}$	$\frac{3}{30}$	$\frac{3}{30}$	$\frac{4}{30}$	$\frac{1}{30}$	$\frac{1}{30}$

$\mu = 5.00467 \approx 5.00; \quad V(X) = 0.0009 ; \quad \sigma = \sqrt{0.0009} = 0.03$

7. a.

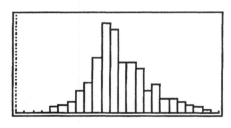

b. $\mu = 65.875$ and $\sigma = 1.73.$ $\qquad\qquad$ 19.5 years old and $\sigma = 2.68.$

EXERCISES 8.4, page 484

1. Yes. The number of trials is fixed, there are two outcomes of the experiment, the probability in each trial is fixed $(p = \frac{1}{6})$, and the trials are independent of each other.

3. No. There are more than 2 outcomes in each trial.

5. No. There are more than 2 outcomes in each trial and the probability of success (an accident) in each trial is not the same.

7. $C(4,2)(\frac{1}{3})^2(\frac{2}{3})^2 = \dfrac{4!}{2!2!}(\frac{4}{81}) \approx 0.296$.

9. $C(5,3)(0.2)^3(0.8)^2 = (\frac{5!}{2!3!})(0.2)^3(0.8)^2 \approx 0.0512$.

11. The required probability is given by $P(X = 0) = C(5,0)(\frac{1}{3})^0(\frac{2}{3})^5 \approx 0.132$.

13. The required probability is given by
$$P(X \geq 3) = C(6,3)(\tfrac{1}{2})^3(\tfrac{1}{2})^{6-3} + C(6,4)(\tfrac{1}{2})^4(\tfrac{1}{2})^{6-4} + C(6,5)(\tfrac{1}{2})^5(\tfrac{1}{2})^{6-5}$$
$$+ C(6,6)(\tfrac{1}{2})^6(\tfrac{1}{2})^{6-6}$$
$$= \tfrac{6!}{3!3!}(\tfrac{1}{2})^6 + \tfrac{6!}{4!2!}(\tfrac{1}{2})^6 + \tfrac{6!}{5!1!}(\tfrac{1}{2})^6 + \tfrac{6!}{6!0!}(\tfrac{1}{2})^6 = \tfrac{1}{64}(20 + 15 + 6 + 1) = \tfrac{21}{32}.$$

15. The probability of no failures, or, equivalently, the probability of five successes is
$$P(X = 5) = C(5,5)(\tfrac{1}{3})^5(\tfrac{2}{3})^{5-5} = \tfrac{1}{243} \approx 0.00412.$$

17. Here $n = 4$, and $p = 1/6$. Then $P(X = 2) = C(4,2)(\frac{1}{6})^2(\frac{5}{6})^2 = (\frac{25}{216}) \approx 0.116$.

19. Here $n = 5$, $p = 0.4$, and therefore, $q = 1 - 0.4 = 0.6$.
   a. $P(X = 0) = C(5,0)(.4)^0(.6)^5 \approx .078$; $P(X = 1) = C(5,1)(.4)(.6)^4 \approx .259$
   $P(X = 2) = C(5,2)(.4)^2(.6)^3 \approx .346$; $P(X = 3) = C(5,3)(.4)^3(.6)^3 \approx .230$
   $P(X = 4) = C(5,4)(.4)^4(.6)^2 \approx .077$; $P(X = 5) = C(5,5)(.4)^5(.6) \approx .010$

b.

X	0	1	2	3	4	5
P(X = x)	0.078	0.259	0.346	0.230	0.077	0.010

c. $\mu = np = 5(0.4) = 2$; $\sigma = \sqrt{npq} = \sqrt{5(0.4)(0.6)} = \sqrt{1.2} \approx 1.095$.

21. Here $1 - p = 1/50$ or $p = 49/50$. So the probability of obtaining 49 or 50 nondefective fuses is
$$P(X = 49) + P(X = 50) = C(50,49)(\tfrac{49}{50})^{49}(\tfrac{1}{50}) + C(50,50)(\tfrac{49}{50})^{50}(\tfrac{1}{50})^{0} \approx 0.74.$$
This is also the probability of at most one defective fuse. So the inference is incorrect.

23. The probability that she will serve 0 or 1 ace is given by
$$P(X = 0) + P(X = 1) = C(5,0)(.15)^{0}(.85)^{5} + C(5,1)(.15)^{1}(.85)^{4}$$
$$= \frac{5!}{0!5!}(.85)^{5} + \frac{5!}{1!4!}(.15)(.85)^{4} \approx 0.83521.$$
Therefore, the probability that she will serve at least two aces is
$$1 - P(X = 0) - P(X = 1) = 1 - .83521 = .16479, \text{ or approximately } .165.$$

25. The required probability is given by $P(X = 6) = C(6,6)(\tfrac{1}{4})^{6}(\tfrac{3}{4})^{0} = 0.0002$.

27. a. The probability that six or more people stated a preference for brand A is
$$P(X \geq 6) = C(10,6)(.6)^{6}(.4)^{4} + C(10,7)(.6)^{7}(.4)^{3}$$
$$+ C(10,8)(.6)^{8}(.4)^{2} + C(10,9)(.6)^{9}(.4)^{1}$$
$$+ C(10,10)(.6)^{10}(.4)^{0}$$
$$\approx 0.251 + 0.215 + 0.121 + 0.040 + 0.006 = 0.633.$$

b. The required probability is $1 - 0.633 = 0.367$.

29. a. Let X denote the number of new buildings that are in violation of the building code. Then $p = \frac{1}{3}$ and $p = \frac{2}{3}$. Therefore, the probability that the first 3 new buildings will pass inspection and the remaining 2 will fail the inspection is $\left(\frac{2}{3}\right)^3 \left(\frac{1}{3}\right)^2 = \frac{2^3}{3^5}$, or approximately 0.0329.

   b. The probability that just 3 of the new buildings will pass inspection (and so 2 will fail the inspection) is $C(5,2)\left(\frac{2}{3}\right)^3 \left(\frac{1}{3}\right)^2 = \frac{5!}{3!2!} \cdot \frac{2^3}{3^5}$, or approximately 0.3292.

31. This is a binomial experiment with $n = 9$, $p = 1/3$, and $q = 2/3$.
   a. The probability is given by
   $$P(X = 3) = C(9,3)(\tfrac{1}{3})^3(\tfrac{2}{3})^6 = (\tfrac{9!}{6!3!})(\tfrac{1}{3})^3(\tfrac{2}{3})^6 \approx 0.273.$$
   b. The probability is given by
   $$P(X = 0) + P(X = 1) + P(X = 2) + P(X = 3)$$
   $$= C(9,0)(\tfrac{1}{3})^0(\tfrac{2}{3})^9 + C(9,1)(\tfrac{1}{3})(\tfrac{2}{3})^8 + C(9,2)(\tfrac{1}{3})^2(\tfrac{2}{3})^7 + C(9,3)(\tfrac{1}{3})^3(\tfrac{2}{3})^6$$
   $$= 0.026 + 0.117 + 0.234 + 0.273 \approx 0.650.$$

33. This is a binomial experiment with $n = 10$, $p = 0.02$, and $q = 0.98$.
   a. The probability that the sample contains no defectives is given by
   $$P(X = 0) = C(10,0)(0.02)^0(0.98)^{10} \approx 0.817.$$
   b. The probability that the sample contains at most 2 defective sets is given by
   $$P(X \leq 2) = C(10,0)(0.02)^0(0.98)^{10} + C(10,1)((0.02)(0.98)^9$$
   $$+ C(10,2)(0.02)^2(0.98)^8$$
   $$\approx 0.817 + 0.167 + 0.015 = 0.999.$$

35. a. The required probability is $P(X = 2) = C(10,2)(0.05)^2(0.95)^8 \approx 0.075$.
   b. The required probability is
   $$P(X \geq 2) = 1 - P(X \leq 2)$$
   $$= 1 - [C(10,2)(.05)^2(.95)^8 + C(10,1)(.05)(.95)^9$$
   $$+ C(10,0)(.05)^0(.95)^{10}]$$
   $$\approx 0.012.$$

37. a. The required probability is
   $$P(X = 0) = C(20,0)(0.1)^0(0.9)^{20} \approx 0.1216.$$

b. The required probability is
$$P(X = 0) = C(20,0)(0.05)^0(0.95)^{20} \approx 0.3585.$$

39. The required probability is $P(X = 0) = C(10,0)(0.1)^0(0.9)^{10} \approx 0.3487.$

41. The mean number of people for whom the drug is effective is
$$\mu = np = (500)(0.75) = 375.$$
The standard deviation of the number of people for whom the drug can be expected to be effective is $\sigma = \sqrt{npq} = \sqrt{(500)(0.75)(0.25)} \approx 9.68.$

43. False. There are exactly two outcomes.

45. False. Here $p = \frac{1}{4}$ and $q = 1 - \frac{1}{4} = \frac{3}{4}$. The probability that the batter will get a hit if he bats four times is
$$1 - P(X = 0) = 1 - C(4,0)(\tfrac{1}{4})^0(\tfrac{3}{4})^4 = 1 - \frac{4!}{4!0!} \cdot 1 \cdot \frac{81}{256} = 1 - \frac{81}{256} = \frac{175}{256} \approx 0.68,$$
which is far from guaranteeing that he gets a hit.

EXERCISES 8.5, page 494

1. $P(Z < 1.45) = 0.9265$

3. $P(Z < -1.75) = 0.0401.$

5. $P(-1.32 < Z < 1.74) = P(Z < 1.74) - P(Z < -1.32) = 0.9591 - 0.0934 = 0.8657.$

7. $P(Z < 1.37) = 0.9147.$

9. $P(Z < -0.65) = 0.2578.$

11. $P(Z > -1.25) = 1 - P(Z < -1.25) = 1 - 0.1056 = 0.8944$

13. $P(0.68 < Z < 2.02) = P(Z < 2.02) - P(Z < 0.68)$
$= 0.9783 - 0.7517 = 0.2266.$

15. a. Referring to Table 2, we see that $P(Z < z) = 0.8907$ implies that $z = 1.23$.
b. Referring to Table 2, we see that $P(Z < z) = 0.2090$ implies that $z = -0.81$.

17. a. $P(Z > -z) = 1 - P(Z < -z) = 1 - 0.9713 = 0.0287$ implies $z = 1.9$.
b. $P(Z < -z) = 0.9713$ implies that $z = -1.9$.

19. a. $P(X < 60) = P(Z < \dfrac{60 - 50}{5}) = P(Z < 2) = 0.9772.$

b. $P(X > 43) = P(Z > \dfrac{43 - 50}{5}) = P(Z > -1.4) = P(Z < 1.4) = 0.9192.$

c. $P(46 < X < 58) = P(\dfrac{46 - 50}{5} < Z < \dfrac{58 - 50}{5}) = P(-0.8 < Z < 1.6)$
$= P(Z < 1.6) - P(Z < -0.8) = 0.9452 - 0.2119 = 0.7333.$

EXERCISES 8.6, page 501

1. $\mu = 20$ and $\sigma = 2.6$.

a. $P(X > 22) = P(Z > \dfrac{22 - 20}{2.6}) = P(Z > 0.77) = P(Z < -0.77) = 0.2206.$

b. $P(X < 18) = P(Z < \dfrac{18 - 20}{2.6}) = P(Z < -0.77) = 0.2206.$

c. $P(19 < X < 21) = P(\dfrac{19 - 20}{2.6} < Z < \dfrac{21 - 20}{2.6}) = P(-0.38 < Z < 0.38)$

$$= P(Z < 0.38) - P(Z < -0.38) = 0.6480 - 0.3520 = 0.2960.$$

3. $\mu = 750$ and $\sigma = 75$.

a. $P(X > 900) = P(Z > \dfrac{900 - 750}{75}) = P(Z > 2) = P(Z < -2) = 0.0228.$

b. $P(X < 600) = P(Z < \dfrac{600 - 750}{75}) = P(Z < -2) = 0.0228.$

c. $P(750 < X < 900) = P(Z < \dfrac{750 - 750}{75} < Z < \dfrac{900 - 750}{75})$
$$= P(0 < Z < 2) = P(Z < 2) - P(Z < 0)$$
$$= 0.9772 - 0.5000 = 0.4772.$$

d. $P(600 < X < 800) = P(\dfrac{600 - 750}{75} < Z < \dfrac{800 - 750}{75})$
$$= P(-2 < Z < .67) = P(Z < .67) - P(Z < -2)$$
$$= 0.7486 - 0.0228 = 0.7258.$$

5. $\mu = 100$ and $\sigma = 15$.

a. $P(X > 140) = P(Z > \dfrac{140 - 100}{15}) = P(Z > 2.67) = P(Z < -2.67) = 0.0038.$

b. $P(X > 120) = P(Z > \dfrac{120 - 100}{15}) = P(Z > 1.33) = P(Z < -1.33) = 0.0918.$

c. $P(100 < X < 120) = P(\dfrac{100 - 100}{15} < Z < \dfrac{120 - 100}{15}) = P(0 < Z < 1.33)$
$$= P(Z < 0) - P(Z < 1.33) = 0.9082 - 0.5000 = 0.4082.$$

d. $P(X < 90) = P(Z < \dfrac{90 - 100}{15}) = P(Z < -0.67) = 0.2514.$

7. Here $\mu = 575$ and $\sigma = 50$.

$P(550 < X < 650) = P(\dfrac{550 - 575}{50} < Z < \dfrac{650 - 575}{50}) = P(-0.5 < Z < 1.5)$
$$= P(Z < 1.5) - P(Z < -0.5) = 0.9332 - 0.3085 = 0.6247.$$

9. Here $\mu = 22$ and $\sigma = 4$.

$P(X < 12) = P(Z < \dfrac{12 - 22}{4}) = P(Z < -2.5) = 0.0062,$ or 0.62 percent.

11. $\mu = 70$ and $\sigma = 10$.
To find the cut-off point for an A, we solve $P(Y < y) = 0.85$ for y. Now

$$P(Y < y) = P\left(Z < \frac{y - 70}{10}\right) = 0.85 \text{ implies } \frac{y - 70}{10} = 1.04, \text{ or } y = 80.4 \approx 80.$$

For a B: $P(Y < y) = P\left(Z < \frac{y - 70}{10}\right) = 0.75$ implies $\frac{y - 70}{10} = 0.67$, or $y \approx 77$.

For a C: $P(Y < y) = P\left(Z < \frac{y - 70}{10}\right) = 0.60$ implies $\frac{y - 70}{10} = 0.25$, or $y \approx 73$.

For a D: $P\left(Z \leq \frac{y - 70}{10}\right) = 0.2$ implies $\frac{y - 70}{10} = -0.84$ or $y \approx 62$.

For a F: $P\left(Z < \frac{y - 70}{10}\right) = 0.05$ implies $\frac{y - 70}{10} = -1.65$, or $y \approx 54$.

13. Let X denote the number of heads in 25 tosses of the coin. Then X is a binomial random variable. Also, $n = 25$, $p = 0.4$, and $q = 0.6$. So
$$\mu = (25)(0.4) = 10; \quad \sigma = \sqrt{(25)(0.4)(0.6)} \approx 2.45.$$

Approximating the binomial distribution by a normal distribution with a mean of 10 and a standard deviation of 2.45, we find upon letting Y denote the associated normal random variable,

a. $P(X < 10) \approx P(Y < 9.5) = P\left(Z < \frac{9.5 - 10}{2.45}\right) = P(Z < -0.20) = 0.4207.$

b. $P(10 \leq X \leq 12) \approx P(9.5 < Y < 12.5)$

$\qquad = P\left(\frac{9.5 - 10}{2.45} < Z < \frac{12.5 - 10}{2.45}\right) = P(Z < 1.02) - P(Z < -0.20)$

$\qquad = P(Z < 1.02) - P(Z < -0.20) = 0.8461 - 0.4207 = 0.4254.$

c. $P(X > 15) \approx P(Y \geq 15)$

$\qquad = P(Z > \frac{15.5 - 10}{2.45}) = P(Z > 2.24) = P(Z < -2.24) = 0.0125.$

15. Let X denote the number of times the marksman hits his target. Then X has a binomial distribution with $n = 30$, $p = 0.6$ and $q = 0.4$. Therefore,
$$\mu = (30)(0.6) = 18, \quad \sigma = \sqrt{(30)(0.6)(0.4)} = 2.68.$$
a. $P(X \geq 20) \approx P(Y \geq 19.5)$

$$= P\left(Z > \frac{19.5 - 18}{2.68}\right) = P(Z > 0.56) = P(Z < -0.56) = 0.2877.$$

b. $P(X < 10) \approx P(Y < 9.5) = P\left(Z < \frac{9.5 - 18}{2.68}\right) = P(Z < -3.17) = 0.0008.$

c. $P(15 \le X \le 20) \approx P(14.5 < Y < 20.5) = P\left(\frac{14.5 - 18}{2.68} < Z < \frac{20.5 - 18}{2.68}\right)$

$$= P(Z < 0.93) - P(Z < -1.31) = 0.8238 - 0.0951 = 0.7287.$$

17.   Let X denote the number of "seconds." Then X has a binomial distribution with $n = 200$, $p = 0.03$, and $q = 0.97$. Then

$$\mu = (200)(0.03) = 6; \quad \sigma = \sqrt{(200)(0.03)(0.97)} \approx 2.41,$$

and $P(X < 10) \approx P(Y < 9.5) = P(Z < \frac{9.5 - 6}{2.41}) = P(Z < 1.45) = 0.9265.$

19.   Let X denote the number of workers who meet with an accident during a 1-year period. Then $\mu = (800)(0.1) = 80$; $\sigma = \sqrt{(800)(0.1)(0.9)} \approx 8.49$,

and $P(X > 70) \approx P(Y > 70.5)$

$$= P\left(Z > \frac{70.5 - 80}{8.49}\right) = P(Z > -1.12) = P(Z < 1.12) = 0.8686.$$

21.   a. Let X denote the number of mice that recovered from the disease. Then X has a binomial distribution with $n = 50$, $p = 0.5$, and $q = 0.5$, so

$$\mu = (50)(0.5) = 25; \quad \sigma = \sqrt{(50)(0.5)(0.5)} \approx 3.54,$$

Approximating the binomial distribution by a normal distribution with a mean of 25 and a standard deviation of 3.54, we find that the probability that 35 or more of the mice would recover from the disease without benefit of the drug is

$$P(X \ge 35) \approx P(Y > 34.5)$$

$$= P(Z > \frac{34.5 - 25}{3.54}) = P(Z > 2.68) = P(Z < -2.68) = 0.0037.$$

b. The drug is very effective.

23.   Let n denote the number of reservations the company should accept. Then we need

to find

$$P(X \geq 2000) \approx P(Y > 1999.5) = 0.01$$

or equivalently,

$$P(Z \geq \frac{1999.5 - np}{\sqrt{npq}}) = 0.01 \qquad \text{[Here p = 0.92 and q = 0.08.]}$$

or $\quad P(Z \leq \frac{np - 1999.5}{\sqrt{npq}}) = 0.01$. Next, $\frac{.92n - 1999.5}{\sqrt{0.0736n}} = -2.33$

$$(0.92n - 1999.5)^2 = (-2.33)^2(0.0736n)$$
$$0.8464n^2 - 3679.08n + 3,998,000.25 = 0.39956704n,$$

or $\quad 0.8464n^2 - 3679.479567n + 3,998,000.25 = 0.$

Using the quadratic formula, we obtain $n = \dfrac{3679.479567 \pm \sqrt{2940.2376}}{1.6928} \approx 2142,$

or 2142. [You can verify that 2206 is not a root of the original equation (before squaring).] Therefore, the company should accept no more than 2142 reservations.

CHAPTER 8, REVIEW EXERCISES, page 505

1. a. $S = \{WWW, WWB, WBW, WBB, BWW, BWB, BBW, BBB\}$

    b.

Outcome	WWW	BWW	WBW	WWB	BBW	BWB	WBB	BBB
Value	0	1	1	1	2	2	2	3

    c.

x	0	1	2	3
$P(X = x)$	$\frac{1}{35}$	$\frac{12}{35}$	$\frac{18}{35}$	$\frac{4}{35}$

    d.

3.   a. $P(1 \leq X \leq 4) = 0.1 + 0.2 + 0.3 + 0.2 = 0.8$.
     b. $\mu = 0(0.1) + 1(0.1) + 2(0.2) + 3(0.3) + 4(0.2) + 5(0.1) = 2.7$.
        $V(X) = 0.1(0 - 2.7)^2 + 0.1(1 - 2.7)^2 + 0.2(2 - 2.7)^2 + 0.3(3 - 2.7)^2$
        $\qquad\qquad + 0.2(4 - 2.7)^2 + 0.1(5 - 2.7)^2$
        $\qquad = 2.01$
        $\qquad \sigma = \sqrt{2.01} \approx 1.418$.

5. $P(Z < 0.5) = 0.6915$.

0.05

7. $P(-0.75 < Z < 0.5) = P(Z < 0.5) - P(Z < -0.75) = 0.6915 - 0.2266 = 0.4649$.

$-0.75\ 0\ 0.5$

9.   If $P(Z < z) = 0.9922$, then $z = 2.4211$.

11. If $P(Z > z) = 0.9788$, then $P(Z < -z) = 0.9788$, and $-z = 2.03$, or $z = -2.03$.

13. $P(X < 11) = P\left(Z < \dfrac{11-10}{2} \right) = P(Z < 0.5) = 0.6915$.

15. $P(7 < X < 9) = P\left(\dfrac{7-10}{2} < Z < \dfrac{9-10}{2} \right) = P(-1.5 < Z < -0.5)$
        $\qquad = P(Z < -0.5) - P(Z < -1.5) = 0.3085 - 0.0668 = 0.2417$.

17. This is a binomial experiment with $p = 0.7$, and so $q = 0.3$. The probability that he
     will get exactly two strikes in four attempts is given by
        $P(X = 2) = C(4,2)(0.7)^2(0.3)^2 \approx 0.2646$.

The probability that he will get at least two strikes in four attempts is given by

$$P(X=2) + P(X=3) + P(X=4)$$
$$= C(4,2)(0.7)^2(0.3)^2 + C(4,3)(0.7)^3(0.3) + C(4,4)(0.7)^4(0.3)^0$$
$$= 0.2646 + 0.4116 + 0.2401 \approx 0.9163.$$

19. Here $\mu = 64.5$ and $\sigma = 2.5$. Next, $64.5 - 2.5k = 59.5$ and $64.5 + 2.5k = 69.5$ and $k = 2$. Therefore, the required probability is given by

$$P(59.5 \leq X \leq 69.5) \geq 1 - \frac{1}{2^2} = 0.75.$$

21. Let the random variable X be the number of people for whom the drug is effective. Then $\mu = (0.15)(800) = 120$ and $\sigma = \sqrt{(800)(0.15)(0.85)} = \sqrt{102} \approx 10.1$.

23. Here $\mu = (0.6)(100) = 60$ and $\sigma = \sqrt{100(0.6)(0.4)} = 4.899$.

Then $P(X > 50) \approx P(Y > 50.5) = P\left(Z > \dfrac{50.5 - 60}{4.899} \right) = P(Z > -1.94)$

$$= P(Z < 1.94) = 0.9738.$$

CHAPTER 9

EXERCISES 9.1, page 514

1. Yes. All entries are nonnegative and the sum of the entries in each column is equal to 1.

3. Yes.

5. No. The sum of the entries of the third column is not 1.

7. Yes.   9. No. It is not a square ($n \times n$) matrix.

11. a. The conditional probability that the outcome state 1 will occur given that the outcome state 1 has occurred is 0.3.
    b. 0.7
    c. We compute $X_1 = TX_0 = \begin{bmatrix} .3 & .6 \\ .7 & .4 \end{bmatrix} \begin{bmatrix} .4 \\ .6 \end{bmatrix} = \begin{bmatrix} .48 \\ .52 \end{bmatrix}$.

13. We compute $TX_0 = \begin{bmatrix} .6 & .2 \\ .4 & .8 \end{bmatrix} \begin{bmatrix} .5 \\ .5 \end{bmatrix} = \begin{bmatrix} .4 \\ .6 \end{bmatrix}$

    Thus, after 1 stage of the experiment, the probability of state 1 occurring is 0.4 and the probability of state 2 occurring is 0.6. The tree diagram describing this process follows.

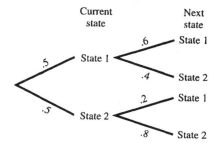

    Using this diagram, we see that the probabilities of state 1 and state 2 occurring in the next stage of the experiment are given by
$$P(S_1) = (0.5)(0.6) + (0.5)(0.2) = 0.4$$

$$P(S_2) = (0.5)(0.4) + (0.5)(0.8) = 0.6$$

Observe that these probabilities are precisely those represented in the probability distribution vector $X_0 T$.

15. $$X_1 = TX_0 = \begin{bmatrix} .4 & .8 \\ .6 & .2 \end{bmatrix}\begin{bmatrix} .6 \\ .4 \end{bmatrix} = \begin{bmatrix} .56 \\ .44 \end{bmatrix}.$$

$$X_2 = TX_1 = \begin{bmatrix} .4 & .6 \\ .6 & .2 \end{bmatrix}\begin{bmatrix} .56 \\ .44 \end{bmatrix} = \begin{bmatrix} .576 \\ .424 \end{bmatrix}.$$

17. $$X_1 = TX_0 = \begin{bmatrix} \frac{1}{4} & \frac{1}{4} & \frac{1}{2} \\ \frac{1}{4} & \frac{1}{2} & \frac{1}{2} \\ \frac{1}{2} & \frac{1}{4} & 0 \end{bmatrix}\begin{bmatrix} \frac{1}{4} \\ \frac{1}{2} \\ \frac{1}{4} \end{bmatrix} = \begin{bmatrix} \frac{5}{16} \\ \frac{7}{16} \\ \frac{1}{4} \end{bmatrix}; \quad X_2 = TX_1 = \begin{bmatrix} \frac{1}{4} & \frac{1}{4} & \frac{1}{2} \\ \frac{1}{4} & \frac{1}{2} & \frac{1}{2} \\ \frac{1}{2} & \frac{1}{4} & 0 \end{bmatrix}\begin{bmatrix} \frac{5}{16} \\ \frac{7}{16} \\ \frac{1}{4} \end{bmatrix} = \begin{bmatrix} \frac{5}{16} \\ \frac{27}{64} \\ \frac{17}{64} \end{bmatrix}.$$

19. a.

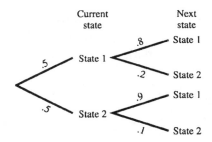

b. $$T = \begin{array}{c} \\ L \\ R \end{array}\begin{array}{cc} L & R \\ \begin{bmatrix} .8 & .9 \\ .2 & .1 \end{bmatrix} \end{array}$$

c. $$X_0 = \begin{array}{c} L \\ R \end{array}\begin{bmatrix} .5 \\ .5 \end{bmatrix}$$

d. $$X_1 = \begin{array}{c} \\ L \\ R \end{array}\begin{array}{cc} L & R \\ \begin{bmatrix} .8 & .9 \\ .2 & .1 \end{bmatrix} \end{array}\begin{bmatrix} .5 \\ .5 \end{bmatrix} = \begin{array}{c} L \\ R \end{array}\begin{bmatrix} .85 \\ .15 \end{bmatrix}$$

21. a. $$X_1 = TX_0 = \begin{array}{c} \\ R \\ D \end{array}\begin{array}{cc} R & D \\ \begin{bmatrix} .7 & .2 \\ .3 & .8 \end{bmatrix} \end{array}\begin{bmatrix} .6 \\ .4 \end{bmatrix} = \begin{array}{c} R \\ D \end{array}\begin{bmatrix} .5 \\ .5 \end{bmatrix}$$

so if the election were held now, it would be a tie.

b.

$$X_1 = TX_0 = \begin{array}{c} R \\ D \end{array}\begin{bmatrix} .7 & .2 \\ .3 & .8 \end{bmatrix}\begin{bmatrix} .5 \\ .5 \end{bmatrix} = \begin{array}{c} R \\ D \end{array}\begin{bmatrix} .45 \\ .55 \end{bmatrix}$$

(with R D column labels above)

so the Democratic candidate would win.

23. After one pickup and discharge the distribution will be

$$X_1 = TX_0 = \begin{bmatrix} .6 & .4 & .3 \\ .3 & .3 & .3 \\ .1 & .3 & .4 \end{bmatrix}\begin{bmatrix} .6 \\ .2 \\ .2 \end{bmatrix} = \begin{bmatrix} .5 \\ .3 \\ .2 \end{bmatrix}$$

or 50% will be in Zone I, 30% will be in Zone II, and 20 percent will be in Zone III.

25. The expected distribution is given by

$$X_1 = TX_0 = \begin{bmatrix} .80 & .10 & .05 \\ .10 & .75 & .05 \\ .10 & .15 & .90 \end{bmatrix}\begin{bmatrix} .4 \\ .4 \\ .2 \end{bmatrix} = \begin{bmatrix} .37 \\ .35 \\ .28 \end{bmatrix}$$

and we conclude that at the beginning of the second quarter the university Bookstore will have 37 percent of the market, the Campus Bookstore will have 35 percent, and the Book Mart will have 28 percent of the market.
Similarly,

$$X_2 = TX_2 = \begin{bmatrix} .80 & .10 & .05 \\ .10 & .75 & .05 \\ .10 & .15 & .90 \end{bmatrix}\begin{bmatrix} .37 \\ .35 \\ .28 \end{bmatrix} = \begin{bmatrix} .345 \\ .3135 \\ .3415 \end{bmatrix}$$

implies that the University Bookstore will have 34.5% of the market, the Campus Bookstore will have 31.35% of the market, and the Book Mart will have 34.15% of the market at the beginning of the third quarter.

27. $$X_1 = TX_0 = \begin{bmatrix} .80 & .10 & .20 & .10 \\ .10 & .70 & .10 & .05 \\ .05 & .10 & .60 & .05 \\ .05 & .10 & .10 & .80 \end{bmatrix}\begin{bmatrix} .3 \\ .3 \\ .2 \\ .2 \end{bmatrix} = \begin{bmatrix} .33 \\ .27 \\ .175 \\ .225 \end{bmatrix}$$

Similarly

$$X_2 = TX_1 = \begin{bmatrix} .3485 \\ .25075 \\ .15975 \\ .241 \end{bmatrix} \quad \text{and} \quad X_3 = TX_2 = \begin{bmatrix} .3599 \\ .2384 \\ .1504 \\ .2513 \end{bmatrix}$$

Assuming that the present trend continues, 36% of the students in their senior year will major in business, 23.8% will major in the humanities, 15% will major in education, and 25.1% will major in the natural sciences.

29. False. In a Markov Chain, an outcome depends only on the preceding stage.

USING TECHNOLOGY EXERCISES 9.1, page 517

1. $$X_5 = \begin{bmatrix} .204489 \\ .131869 \\ .261028 \\ .186814 \\ .2158 \end{bmatrix}$$

3. Manufacturer A will have 23.95% of the market, Manufacturer B will have 49.71% of the market share, and manufacturer C will have 26.34 percent of the market share.

EXERCISES 9.2, page 526

1. Since all entries in the matrix are positive, it is regular.

3. $$T^2 = \begin{bmatrix} 1 & .8 \\ 0 & .2 \end{bmatrix}\begin{bmatrix} 1 & .8 \\ 0 & .2 \end{bmatrix} = \begin{bmatrix} 1 & .96 \\ 0 & .04 \end{bmatrix}; \quad T^3 = \begin{bmatrix} 1 & .96 \\ 0 & .04 \end{bmatrix}\begin{bmatrix} 1 & .8 \\ 0 & .2 \end{bmatrix} = \begin{bmatrix} 1 & .992 \\ 0 & .008 \end{bmatrix}$$

and we see that the a_{21} entry will always be zero, so T is not regular.

5. $$T^2 = \begin{bmatrix} \frac{1}{2} & \frac{3}{4} & 0 \\ \frac{1}{2} & 0 & \frac{1}{2} \\ 0 & \frac{1}{4} & \frac{1}{2} \end{bmatrix}\begin{bmatrix} \frac{1}{2} & \frac{3}{4} & 0 \\ \frac{1}{2} & 0 & \frac{1}{2} \\ 0 & \frac{1}{4} & \frac{1}{2} \end{bmatrix} = \begin{bmatrix} \frac{5}{8} & \frac{3}{8} & \frac{3}{8} \\ \frac{1}{4} & \frac{1}{2} & \frac{1}{4} \\ \frac{1}{8} & \frac{1}{8} & \frac{3}{8} \end{bmatrix}$$

and so this matrix is regular.

7. $T^2 = \begin{bmatrix} .7 & .2 & .3 \\ .3 & .8 & .3 \\ 0 & 0 & .4 \end{bmatrix} \begin{bmatrix} .7 & .2 & .3 \\ .3 & .8 & .3 \\ 0 & 0 & .4 \end{bmatrix} = \begin{bmatrix} .55 & .3 & .39 \\ .45 & .7 & .45 \\ 0 & 0 & .16 \end{bmatrix}$

and so forth.

Continuing, we see that $T^3, T^4, \ldots,$ will have the a_{31} and a_{32} entries equal to zero and T is not regular.

9. We solve the matrix equation

$$\begin{bmatrix} \frac{1}{3} & \frac{1}{4} \\ \frac{2}{3} & \frac{3}{4} \end{bmatrix} \begin{bmatrix} x \\ y \end{bmatrix} = \begin{bmatrix} x \\ y \end{bmatrix}$$

or equivalently, the system of equations

$$\frac{1}{3}x + \frac{1}{4}y = x$$
$$\frac{2}{3}x + \frac{3}{4}y = y$$
$$x + y = 1.$$

Solving this system of equations, we find the required vector to be $\begin{bmatrix} \frac{3}{11} \\ \frac{8}{11} \end{bmatrix}$.

11. We have $TX = X$, that is, $\begin{bmatrix} .5 & .2 \\ .5 & .8 \end{bmatrix} \begin{bmatrix} x \\ y \end{bmatrix} = \begin{bmatrix} x \\ y \end{bmatrix}$

or equivalently, the system of equations

$$.5x + .2y = x$$
$$.5x + .8y = y.$$

These two equations are equivalent to the single equation $0.5x - 0.2y = 0$.

We must also have $x + y = 1$. So we have the system

$$.5x - .2y = x$$
$$x + y = 1$$

The second equation gives $y = 1 - x$, which when substituted into the first equation yields,

$$0.5x - 0.2(1 - x) = 0, \quad 0.7x - 0.2 = 0, \quad \text{or} \quad x = 2/7.$$

Therefore, $y = 5/7$ and the steady-state distribution vector is $\begin{bmatrix} \frac{2}{7} \\ \frac{5}{7} \end{bmatrix}$.

13. We solve the system

$$\begin{bmatrix} 0 & \frac{1}{8} & 1 \\ 1 & \frac{5}{8} & 0 \\ 0 & \frac{1}{4} & 0 \end{bmatrix}\begin{bmatrix} x \\ y \\ z \end{bmatrix} = \begin{bmatrix} x \\ y \\ z \end{bmatrix}$$

together with the equation $x + y + z = 1$; that is, the system

$$-x + \tfrac{1}{8}y + z = 0$$
$$x - \tfrac{3}{8}y \quad\;\; = 0$$
$$\tfrac{1}{4}y - z = 0$$
$$x + y + z = 1.$$

Using the Gauss-Jordan method, we find that the required steady-state vector is

$$\begin{bmatrix} \frac{3}{13} \\ \frac{8}{13} \\ \frac{2}{13} \end{bmatrix}.$$

15. We solve the system

$$\begin{bmatrix} .2 & 0 & .3 \\ 0 & .6 & .4 \\ .8 & .4 & .3 \end{bmatrix}\begin{bmatrix} x \\ y \\ z \end{bmatrix} = \begin{bmatrix} x \\ y \\ z \end{bmatrix}$$

together with the equation $x + y + z = 1$, or equivalently, the system

$$-0.8x \qquad\quad + 0.3z = 0$$
$$-0.4y + 0.4z = 0$$
$$0.8x + 0.4y - 0.7z = 0$$
$$x \;\;+ y \quad + z = 1.$$

Using the Gauss-Jordan method, we find that the required steady-state vector is

$$\begin{bmatrix} \frac{3}{19} \\ \frac{8}{19} \\ \frac{8}{19} \end{bmatrix}.$$

17. a. We want to solve

$$\begin{bmatrix} .8 & .9 \\ .2 & .1 \end{bmatrix}\begin{bmatrix} x \\ y \end{bmatrix} = \begin{bmatrix} x \\ y \end{bmatrix},\ \text{or equivalently}\ \begin{array}{r} .8x + .9y = x \\ .2x + .1y = y \\ x + y = 1 \end{array}$$

Solving this system, we find that the required steady-state vector is $\begin{bmatrix} \frac{2}{11} \\ \frac{9}{11} \end{bmatrix}$

and conclude that in the long run, the mouse will turn left 81.8% of the time.

19. We compute

$$X_1 = \begin{bmatrix} .72 & .12 \\ .28 & .88 \end{bmatrix}\begin{bmatrix} .48 \\ .52 \end{bmatrix} = \begin{bmatrix} .408 \\ .592 \end{bmatrix},$$

and conclude that, ten years from now, there will be 40.8 percent 1-wage-earners and 59.2% 2-wage earners.

b. We solve the system
$$\begin{bmatrix} .72 & .12 \\ .28 & .88 \end{bmatrix}\begin{bmatrix} x \\ y \end{bmatrix} = \begin{bmatrix} x \\ y \end{bmatrix}$$
together with the equation $x + y = 1$.
$$\begin{array}{r} -0.28x + 0.12y = 0 \\ 0.28x - 0.12y = 0 \\ x + y = 1 \end{array}$$
Solving, we find $x = 0.3$ and $y = 0.7$, and conclude that in the long run, there will be 30% 1-wage earners and 70% 2-wage earners.

21. a. If this trend continues, the percentage of homeowners in this city who will own single-family homes or condominiums two years from now will be given by $X_2 = TX_1$. Thus,

$$X_1 = TX_0 = \begin{bmatrix} .85 & .35 \\ .15 & .65 \end{bmatrix}\begin{bmatrix} .8 \\ .2 \end{bmatrix} = \begin{bmatrix} .75 \\ .25 \end{bmatrix}$$

$$X_2 = TX_1 = \begin{bmatrix} .85 & .35 \\ .15 & .65 \end{bmatrix}\begin{bmatrix} .75 \\ .25 \end{bmatrix} = \begin{bmatrix} .725 \\ .275 \end{bmatrix}$$

and we conclude that 72.5% will own single-family home and 27.5% will own condominiums at that time.

b. We solve the system

$$\begin{bmatrix} .85 & .35 \\ .15 & .65 \end{bmatrix} \begin{bmatrix} x \\ y \end{bmatrix} = \begin{bmatrix} x \\ y \end{bmatrix}$$

together with the equation $x + y = 1$. Thus,

$$-0.15x + 0.35y = 0$$
$$0.15x - 0.35y = 0$$
$$x + \quad y = 1$$

Solving, we find $x = 0.7$ and $y = 0.3$, and conclude that in the long run 70% will own single family homes and 30% will own condominiums.

23. a.

$$X_1 = TX_0 = \begin{bmatrix} .8 & .1 & .1 \\ .1 & .85 & .05 \\ .1 & .05 & .85 \end{bmatrix} \begin{bmatrix} .3 \\ .4 \\ .3 \end{bmatrix} = \begin{bmatrix} .31 \\ .385 \\ .305 \end{bmatrix}$$

$$X_2 = TX_1 = \begin{bmatrix} .8 & .1 & .1 \\ .1 & .85 & .05 \\ .1 & .05 & .85 \end{bmatrix} \begin{bmatrix} .31 \\ .385 \\ .305 \end{bmatrix} = \begin{bmatrix} .317 \\ .3735 \\ .3095 \end{bmatrix}$$

From our computations, we conclude that after two weeks 31.7% of the viewers will watch the *ABC* news, 37.35% will watch the *CBS* news, and 30.95% will watch the *NBC* news.

b. We solve the system

$$\begin{bmatrix} .8 & .1 & .1 \\ .1 & .85 & .05 \\ .1 & .05 & .85 \end{bmatrix} \begin{bmatrix} x \\ y \\ z \end{bmatrix} = \begin{bmatrix} x \\ y \\ z \end{bmatrix}$$

together with the equation $x + y + z = 1$, or equivalently, the system

$$-0.2x + \quad 0.1y + \quad 0.1z = 0$$
$$0.1x - 0.15y + 0.05z = 0$$
$$0.1x + 0.05y - 0.15z = 0$$
$$x + \quad y + \quad z = 1$$

Using the Gauss-Jordan elimination method, we find that the required steady-state vector is $\begin{bmatrix} \frac{1}{3} \\ \frac{1}{3} \\ \frac{1}{3} \end{bmatrix}$, and conclude that each network will comand 33 1/3 % of the audience

in the long run.

25. We wish to solve

$$\begin{bmatrix} \frac{1}{2} & \frac{1}{4} & 0 \\ \frac{1}{2} & \frac{1}{2} & \frac{1}{2} \\ 0 & \frac{1}{4} & \frac{1}{2} \end{bmatrix}\begin{bmatrix} x \\ y \\ z \end{bmatrix} = \begin{bmatrix} x \\ y \\ z \end{bmatrix}$$

together with the equation $x + y + z = 1$, or, equivalently, the system of equations

$$\frac{1}{2}x + \frac{1}{4}y \qquad = x$$
$$\frac{1}{2}x + \frac{1}{2}y + \frac{1}{2}z = y$$
$$\frac{1}{4}y + \frac{1}{2}z = z$$
$$x + \ y + \ z = 1$$

Solving this system, we find that
$$x = \tfrac{1}{4}, \ y = \tfrac{1}{2}, \text{ and } z = \tfrac{1}{4}.$$

Thus, in the long run, 25% of the plants will have red flowers, 50% will have pink flowers and 25% will have white flowers.

27. False. All the entries of the limiting matrix must be positive as well.

29. Let T be a regular stochastic matrix and X the steady-state distribution vector that satisfies the equation $TX = T$ and assume that the sum of the elements of X are equal to 1. Then $TX = X$ implies that $TX = T^2 X$, or $X = T^2 X$,.... So we have $X = T^n X$. Next, let L be the steady-state distribution vector, then

$$L = \lim_{m \to \infty} X_m = \lim_{m \to \infty} T^m X_0 = T^m X_0$$

when m is large. Multiplying both sides by T, we obtain

$$TL = T^{m+1} X_0 \approx L.$$

Thus, L also satisfies $TL = L$ together with the condition that the sum of the elements in L be equal to 1. Since the matrix equation $TX = X$ has a unique solution, we conclude that $X = L$.

1. $X_5 = \begin{bmatrix} 0.2045 \\ 0.1319 \\ 0.2610 \\ 0.1868 \\ 0.2158 \end{bmatrix}$

EXERCISES 9.3, page 536

1. The given matrix is an absorbing stochastic matrix

$$T = \begin{array}{c} \\ 1 \\ 2 \end{array}\begin{array}{cc} 1 & 2 \\ \begin{bmatrix} \frac{2}{5} & 0 \\ \frac{3}{5} & 1 \end{bmatrix} \end{array}$$

State 2 is an absorbing state. State 1 is nonabsorbing, but an object in this state has a probability of 3/5 of going to the absorbing state 2.

3. The given matrix is

$$\begin{array}{c} \\ 1 \\ 2 \\ 3 \end{array}\begin{array}{ccc} 1 & 2 & 3 \\ \begin{bmatrix} 1 & .5 & 0 \\ 0 & 0 & 1 \\ 0 & .5 & 0 \end{bmatrix} \end{array}$$

States 1 and 3 are absorbing states. State 2 is not absorbing, but an object in this state has a probability of .5 of going to the absorbing state 1 and .5 of going to the absorbing state 3. Thus, the matrix is an absorbing matrix.

5. Yes. It is an absorbing stochastic matrix since it is possible to go from state 1 to the absorbing states 2 and 3.

7. The given matrix is

$$\begin{array}{c c c c c}
 & 1 & 2 & 3 & 4 \\
1 & \begin{bmatrix} 1 \\ 0 \\ 0 \\ 0 \end{bmatrix} & \begin{matrix} 0 \\ 1 \\ 0 \\ 0 \end{matrix} & \begin{matrix} .3 \\ .2 \\ .1 \\ .4 \end{matrix} & \begin{matrix} 0 \\ 0 \\ .5 \\ .5 \end{bmatrix}
\end{array}$$

States 1 and 2 are absorbing states. States 3 and 4 are not. However, it is possible for an object to go from state 3 to state 1 with probability 0.3. furthermore, it is also possible for an object to go from the non-absorbing state 4 to an absorbing state. For example, via state 3 with a probability of 0.5. Therefore, the given matrix is an absorbing matrix.

9. The required matrix is
$$\begin{array}{c c c}
 & 2 & 1 \\
2 & \begin{bmatrix} 1 & .4 \\ 0 & .6 \end{bmatrix} \\
1
\end{array}$$
where $S = [.4]$ and $R = [.6]$

11.
$$\begin{array}{c c c c}
 & 3 & 2 & 1 \\
3 & \begin{bmatrix} 1 & .4 & .5 \\ 0 & .4 & .5 \\ 0 & .2 & 0 \end{bmatrix} \\
2 \\
1
\end{array}$$
where $S = \begin{bmatrix} .4 & .5 \end{bmatrix}$ and $R = \begin{bmatrix} .4 & .5 \\ .2 & 0 \end{bmatrix}$

or
$$\begin{array}{c c c c}
 & 3 & 1 & 2 \\
3 & \begin{bmatrix} 1 & .5 & .4 \\ 0 & 0 & .2 \\ 0 & .5 & .4 \end{bmatrix} \\
1 \\
2
\end{array}$$
where $S = \begin{bmatrix} .5 & .4 \end{bmatrix}$ and $R = \begin{bmatrix} 0 & .2 \\ .5 & .4 \end{bmatrix}$.

13.
$$\begin{bmatrix} 1 & 0 & .2 & .4 \\ 0 & 1 & .3 & 0 \\ 0 & 0 & .3 & .2 \\ 0 & 0 & .2 & .4 \end{bmatrix} , \quad S = \begin{bmatrix} .2 & .4 \\ .3 & 0 \end{bmatrix}, \quad R = \begin{bmatrix} .3 & .2 \\ .2 & .4 \end{bmatrix}$$

$$\text{or } \begin{bmatrix} 1 & 0 & \vdots & .4 & .2 \\ 0 & 1 & \vdots & 0 & .3 \\ \cdots & \cdots & \vdots & \cdots & \cdots \\ 0 & 0 & \vdots & .4 & .2 \\ 0 & 0 & \vdots & .2 & .3 \end{bmatrix}, \quad S = \begin{bmatrix} .4 & .2 \\ 0 & .3 \end{bmatrix}, \quad R = \begin{bmatrix} .4 & .2 \\ .2 & .3 \end{bmatrix} \quad \text{and so forth.}$$

15. Rewriting the matrix so that the absorbing states appear first, we have

$$\begin{array}{cc} & \begin{array}{cc} 2 & 1 \end{array} \\ \begin{array}{c} 2 \\ 1 \end{array} & \begin{bmatrix} 1 & \vdots & .45 \\ \cdots & \vdots & \cdots \\ 0 & \vdots & .55 \end{bmatrix} \end{array} \quad \text{where } S = [.45] \text{ and } R = [.55]. \quad \text{Then}$$

$$(I - R) = [.45] \quad \text{and} \quad (I - R)^{-1} = [.45]\left[\frac{1}{.45}\right] = 1$$

Therefore the steady-state matrix is $\begin{array}{cc} & \begin{array}{cc} 2 & 1 \end{array} \\ \begin{array}{c} 2 \\ 1 \end{array} & \begin{bmatrix} 1 & \vdots & 1 \\ \cdots & \vdots & \cdots \\ 0 & \vdots & 0 \end{bmatrix} \end{array}.$

17. Here we have $\begin{bmatrix} 1 & \vdots & .2 & .3 \\ \cdots & \vdots & \cdots & \cdots \\ 0 & \vdots & .4 & .2 \\ 0 & \vdots & .4 & .5 \end{bmatrix}$ where $S = [.2 \quad .3]$ and $R = \begin{bmatrix} .4 & .4 \\ .2 & .5 \end{bmatrix}$

Next, $I - R = \begin{bmatrix} 1 & 0 \\ 0 & 1 \end{bmatrix} - \begin{bmatrix} .4 & .2 \\ .4 & .5 \end{bmatrix} = \begin{bmatrix} .6 & -.2 \\ -.4 & .5 \end{bmatrix}$

Using the formula for finding the inverse of a 2×2 matrix, we have

$$(I - R)^{-1} = \begin{bmatrix} 2.27 & .91 \\ 1.8 & 2.73 \end{bmatrix}$$

Then $S(I - R)^{-1} = [.2 \quad .3]\begin{bmatrix} 2.27 & .91 \\ 1.8 & 2.73 \end{bmatrix} = [.994 \quad 1] \approx [1 \quad 1]$

We conclude that the steady-state matrix is $\begin{bmatrix} 1 & \vdots & 1 & 1 \\ \cdots & \vdots & \cdots & \cdots \\ 0 & \vdots & 0 & 0 \\ 0 & \vdots & 0 & 0 \end{bmatrix}.$

19. Upon rewriting the given matrix so that the absorbing states appear first, we have

$$
\begin{array}{c}
\;\;2\;\;\;4\;\;\;1\;\;\;3\\
\begin{array}{c}2\\4\\1\\3\end{array}
\left[\begin{array}{cc|cc}
1 & 0 & \frac{1}{2} & 0\\
0 & 1 & 0 & 0\\
\hline
0 & 0 & \frac{1}{2} & \frac{1}{3}\\
0 & 0 & 0 & \frac{2}{3}
\end{array}\right]
\end{array}
$$

where $S = \begin{bmatrix}\frac{1}{2} & 0\\ 0 & 0\end{bmatrix}$ and $R = \begin{bmatrix}\frac{1}{2} & \frac{1}{3}\\ 0 & \frac{2}{3}\end{bmatrix}$. Next, we compute

$$
I - R = \begin{bmatrix}1 & 0\\ 0 & 1\end{bmatrix} - \begin{bmatrix}\frac{1}{2} & \frac{1}{3}\\ 0 & \frac{2}{3}\end{bmatrix} = \begin{bmatrix}\frac{1}{2} & -\frac{1}{3}\\ 0 & \frac{1}{3}\end{bmatrix}.
$$

Using the formula for finding the inverse of a 2×2 matrix, we have

$$
(I - R)^{-1} = \begin{bmatrix}2 & 2\\ 0 & 3\end{bmatrix} \text{ and so } S(I - R)^{-1} = \begin{bmatrix}2 & 2\\ 0 & 3\end{bmatrix}\begin{bmatrix}\frac{1}{2} & 0\\ 0 & 0\end{bmatrix} = \begin{bmatrix}1 & 1\\ 0 & 0\end{bmatrix}.
$$

Therefore, the steady-state matrix is $\begin{bmatrix}1 & 0 & 1 & 1\\ 0 & 1 & 0 & 0\\ \hline 0 & 0 & 0 & 0\\ 0 & 0 & 0 & 0\end{bmatrix}$.

21. Here $\begin{bmatrix}1 & 0 & \frac{1}{4} & \frac{1}{3}\\ 0 & 1 & \frac{1}{4} & \frac{1}{3}\\ \hline 0 & 0 & \frac{1}{2} & 0\\ 0 & 0 & 0 & \frac{1}{3}\end{bmatrix}$, $S = \begin{bmatrix}\frac{1}{4} & \frac{1}{3}\\ \frac{1}{4} & \frac{1}{3}\end{bmatrix}$, $R = \begin{bmatrix}\frac{1}{2} & 0\\ 0 & \frac{1}{3}\end{bmatrix}$

and $I - R = \begin{bmatrix}1 & 0\\ 0 & 1\end{bmatrix} - \begin{bmatrix}\frac{1}{2} & 0\\ 0 & \frac{1}{3}\end{bmatrix} = \begin{bmatrix}\frac{1}{2} & 0\\ 0 & \frac{2}{3}\end{bmatrix}$.

Using the formula for finding the inverse of a 2×2 matrix, we find

$$(I-R)^{-1} = \begin{bmatrix} 2 & 0 \\ 0 & \frac{3}{2} \end{bmatrix} \quad \text{and} \quad S(I-R)^{-1} = \begin{bmatrix} \frac{1}{4} & \frac{1}{3} \\ \frac{1}{4} & \frac{1}{3} \end{bmatrix} \begin{bmatrix} 2 & 0 \\ 0 & \frac{3}{2} \end{bmatrix} = \begin{bmatrix} \frac{1}{2} & \frac{1}{2} \\ \frac{1}{2} & \frac{1}{2} \end{bmatrix}.$$

The steady-state matrix is given by $\begin{bmatrix} 1 & 0 & \frac{1}{2} & \frac{1}{2} \\ 0 & 1 & \frac{1}{2} & \frac{1}{2} \\ 0 & 0 & 0 & 0 \\ 0 & 0 & 0 & 0 \end{bmatrix}.$

23. The absorbing states already appear first in the matrix, so it need not be
rewritten. Next, $(I-R) = \begin{bmatrix} .8 & -.2 \\ -.2 & .6 \end{bmatrix}$ and $(I-R)^{-1} = \begin{bmatrix} \frac{15}{11} & \frac{5}{11} \\ \frac{5}{11} & \frac{20}{11} \end{bmatrix}$ so that

$$S(I-R)^{-1} = \begin{bmatrix} \frac{2}{10} & \frac{1}{10} \\ \frac{1}{10} & \frac{2}{10} \\ \frac{3}{10} & \frac{1}{10} \end{bmatrix} \begin{bmatrix} \frac{15}{11} & \frac{5}{11} \\ \frac{5}{11} & \frac{20}{11} \end{bmatrix} = \begin{bmatrix} \frac{7}{22} & \frac{3}{11} \\ \frac{5}{22} & \frac{9}{22} \\ \frac{5}{11} & \frac{7}{22} \end{bmatrix}$$

Therefore, the steady-state matrix is given by

$$\begin{bmatrix} 1 & 0 & 0 & \frac{7}{22} & \frac{3}{11} \\ 0 & 1 & 0 & \frac{5}{22} & \frac{9}{22} \\ 0 & 0 & 1 & \frac{5}{11} & \frac{7}{22} \\ 0 & 0 & 0 & 0 & 0 \\ 0 & 0 & 0 & 0 & 0 \end{bmatrix}.$$

25. a. State 2 is absorbing. State 1 is not absorbing, but it is possible for an object to
go from state 1 to state 2 with probability 0.8. Therefore, the matrix is
absorbing. Rewriting, we obtain

$$\begin{array}{c} \\ 1 \\ 2 \end{array} \begin{array}{cc} 2 & 1 \end{array} \begin{bmatrix} 1 & .2 \\ 0 & .8 \end{bmatrix} \quad \text{where } S = \begin{bmatrix} 0.2 \end{bmatrix} \text{ and } R = \begin{bmatrix} 0.8 \end{bmatrix}.$$

b. We compute $I - R = \begin{bmatrix} 1 \end{bmatrix} - \begin{bmatrix} 0.2 \end{bmatrix} = \begin{bmatrix} 0.8 \end{bmatrix}$. So $(I-R)^{-1} = \begin{bmatrix} 1.25 \end{bmatrix}$. Therefore,

$S(I-R)^{-1} = [.8][1.25] = [1]$ and the steady state matrix is $\begin{bmatrix} 1 & 1 \\ 0 & 0 \end{bmatrix}$

this result tells us that in the long run only unleaded gas will be used.

27. Here

$$\begin{array}{c} \begin{array}{ccccc} \$0 & \$4 & \$1 & \$2 & \$3 \end{array} \\ \begin{array}{c} \$0 \\ \$4 \\ \$1 \\ \$2 \\ \$3 \end{array} \begin{bmatrix} 1 & 0 & \frac{1}{2} & 0 & 0 \\ 0 & 1 & 0 & 0 & \frac{1}{2} \\ 0 & 0 & 0 & \frac{1}{2} & 0 \\ 0 & 0 & \frac{1}{2} & 0 & \frac{1}{2} \\ 0 & 0 & 0 & \frac{1}{2} & 0 \end{bmatrix} \end{array}$$ where $S = \begin{bmatrix} \frac{1}{2} & 0 & 0 \\ 0 & 0 & \frac{1}{2} \end{bmatrix}$ and $R = \begin{bmatrix} 0 & \frac{1}{2} & 0 \\ \frac{1}{2} & 0 & \frac{1}{2} \\ 0 & \frac{1}{2} & 0 \end{bmatrix}$.

Next, $I-R = \begin{bmatrix} 1 & -\frac{1}{2} & 0 \\ -\frac{1}{2} & 1 & -\frac{1}{2} \\ 0 & -\frac{1}{2} & 1 \end{bmatrix}$ and $(I-R)^{-1} = \begin{bmatrix} \frac{3}{2} & 1 & \frac{3}{2} \\ 1 & 2 & 1 \\ \frac{3}{2} & 1 & \frac{3}{2} \end{bmatrix}$ and

$S(I-R)^{-1} = \begin{bmatrix} \frac{3}{4} & \frac{1}{2} & \frac{1}{4} \\ \frac{1}{4} & \frac{1}{2} & \frac{3}{4} \end{bmatrix}$.

Therefore, the steady-state matrix is given by $\begin{bmatrix} 1 & 0 & \frac{3}{4} & \frac{1}{2} & \frac{1}{4} \\ 0 & 1 & \frac{1}{4} & \frac{1}{2} & \frac{3}{4} \\ 0 & 0 & 0 & 0 & 0 \\ 0 & 0 & 0 & 0 & 0 \\ 0 & 0 & 0 & 0 & 0 \end{bmatrix}$

We conclude that if Diane started out with \$1, the probability that she would leave the game a winner is 1/4. Similarly, if she started out with \$2, the probability that she would leave the game a winner is 1/2, and if she started out with \$3, the probability that she would leave as a winner is 3/4.

29. a.
$$\begin{bmatrix} 1 & 0 & .25 & .1 \\ 0 & 1 & 0 & .9 \\ 0 & 0 & 0 & 0 \\ 0 & 0 & .75 & 0 \end{bmatrix}$$

b. $I - R = \begin{bmatrix} 1 & 0 \\ -.75 & 1 \end{bmatrix}$ and $(I - R)^{-1} = \begin{bmatrix} 1 & 0 \\ .75 & 1 \end{bmatrix}$ and

$$S(I - R)^{-1} = \begin{bmatrix} .25 & .1 \\ 0 & .9 \end{bmatrix}\begin{bmatrix} 1 & 0 \\ .75 & 1 \end{bmatrix} = \begin{bmatrix} .325 & .1 \\ .675 & .9 \end{bmatrix}.$$

Therefore, the steady-state matrix is

$$\begin{bmatrix} 1 & 0 & .325 & .1 \\ 0 & 1 & .675 & .9 \\ \hline 0 & 0 & 0 & 0 \\ 0 & 0 & 0 & 0 \end{bmatrix}$$

c. From the steady-state matrix, we see that the probability that a beginning student enrolled in the program will compete the course successfullly is 0.675.

31. False. It must be possible to go from any nonabsorbing state to an absorbing state in one or more stages.

33. The transition matrix is
$$\begin{array}{c} \\ aa \\ Aa \\ AA \end{array}\begin{array}{c} aa\ \ Aa\ \ AA \\ \begin{bmatrix} 1 & \frac{1}{2} & 0 \\ \hline 0 & \frac{1}{2} & 1 \\ 0 & 0 & 0 \end{bmatrix} \end{array}$$. Since the entries in T are exactly the

same as those in Example 4, the steady-state matrix is

$$\begin{array}{c} \\ aa \\ Aa \\ AA \end{array}\begin{array}{c} aa\ \ Aa\ \ AA \\ \begin{bmatrix} 1 & 1 & 1 \\ \hline 0 & 0 & 0 \\ 0 & 0 & 0 \end{bmatrix} \end{array}$$

Interpreting the steady-state matrix, we see that in the long run all the flowers produced by the plants will be white.

EXERCISES 9.4, page 547

1. We first determine the minimum of each row and the maxima of each column of the payoff matrix. Next, we find the larger of the row minima and the smaller of the column maxima as shown in the following matrix.

C's move
C_1 C_2

R's move
$$\begin{array}{c} R_1 \\ R_2 \end{array} \begin{bmatrix} 2 & 3 \\ 4 & 1 \end{bmatrix} \begin{array}{l} 2 \leftarrow \text{larger of the row minima} \\ 1 \end{array}$$

4 3

↑_____ smaller of the column maxima

From the above results, we conclude that the row player's maximum strategy is to play row 1, whereas the column player's minimax strategy is to play column 2.

3. We first obtain the following matrix, where the larger of the row minima and the smallest of the column maxima are displayed.

C_1 C_2 C_3

$$\begin{array}{c} R_1 \\ R_2 \end{array} \begin{bmatrix} 1 & 3 & 2 \\ 0 & -1 & 4 \end{bmatrix} \begin{array}{l} 1 \leftarrow \text{larger of the row minima} \\ -1 \end{array}$$

1   3   4

↑____ smallest of the column maxima

We conclude that the row player's maximum strategy is to play row 1, whereas the column player's minimax strategy is to play column 1.

5. From the following payoff matrix where the largest of the row minima and the smallest of the column maxima are displayed

$$\begin{bmatrix} 3 & 2 & 1 \\ 1 & -2 & 3 \\ 6 & 4 & 1 \end{bmatrix} \begin{array}{l} 1 \leftarrow \\ -2 \\ 1 \leftarrow \end{array} \text{largest of the row minima}$$

6   4   3

↑____ smallest of the column maxima

we conclude that the row player's maximin strategy is to play either row 1 or row 3, whereas the column player's minimax strategy is to play column 3.

7. From the following payoff matrix, where the largest of the row minima and the smallest of the column maxima are displayed,

$$\begin{bmatrix} 4 & 2 & 1 \\ 1 & 0 & -1 \\ 2 & 1 & 3 \end{bmatrix} \begin{matrix} 1 \leftarrow \\ -1 \\ 1 \leftarrow \end{matrix}$$ — largest of the row minima

4 2 3

↑_____ smallest of the column maxima

we conclude that the row player's maximin strategy is to play either row 1 or row 3, whereas the column player's minimax strategy is to play column 2.

9. From

$$\begin{bmatrix} \boxed{2} & 3 \\ 1 & -4 \end{bmatrix} \begin{matrix} 2 \leftarrow \\ -4 \end{matrix}$$ larger of the row minima

2 3

↑_____ smallest of the column maxima

we see that the game is strictly determined, and
a. the saddle point is 2.
b. the optimum strategy for the row player is to play row 1, whereas the optimum strategy for the column player is to play column 1.
c. the value of the game is 2.
d. the game favors the row player.

11. From

$$\begin{bmatrix} \boxed{1} & 3 & 2 \\ -1 & 4 & -6 \end{bmatrix} \begin{matrix} 1 \leftarrow \\ -6 \end{matrix}$$ larger of the row minima

1 4 2

↑_____ smallest of the column maxima

we see that the game is strictly determined, and
a. the saddle point is 1.
b. the optimum strategy for the row player is to play row 1, whereas, the optimum strategy for the column player is to play column 1.
c. the value of the game is 1 and d. the game favors the row player.

13. From

$$\begin{bmatrix} \boxed{1} & 3 & 4 & 2 \\ 0 & 2 & 6 & -4 \\ -1 & -3 & -2 & -1 \end{bmatrix} \begin{matrix} 1 \leftarrow \text{largest of the row minima} \\ -4 \\ -3 \end{matrix}$$

$$\begin{matrix} \quad 1 \quad\; 3 \quad\; 6 \quad\; 2 \\ \uparrow \underline{\hspace{4cm}} \text{ smallest of the column maxima} \end{matrix}$$

we conclude that the game is strictly determined, and
a. the saddle point is 1.
b. the optimum strategy for the row player is to play row 1, while the optimum strategy for the column player is to play column 1.
c. the value of the game is 1.　　d. the game favors the row player.

15. From

$$\begin{bmatrix} 1 & 2 \\ 0 & 3 \\ -1 & 2 \\ 2 & -2 \end{bmatrix} \begin{matrix} 1 \leftarrow \text{larger of the row minima} \\ 0 \\ -1 \\ -2 \end{matrix}$$

$$\begin{matrix} \quad 2 \quad\; 3 \\ \uparrow \underline{\hspace{2cm}} \text{ smaller of the column maxima} \end{matrix}$$

we see that the game is not strictly determined and consequently has no saddle point.

17. From

$$\begin{bmatrix} 1 & -1 & 3 & 2 \\ 1 & 0 & 2 & 2 \\ -2 & 2 & 3 & -1 \end{bmatrix} \begin{matrix} -1 \\ 0 \leftarrow \text{largest of the row minima} \\ -2 \end{matrix}$$

$$\begin{matrix} \quad 1 \quad\; 2 \quad 3 \quad\; 2 \\ \uparrow \underline{\hspace{4cm}} \text{ smallest of the column maxima} \end{matrix}$$

we conclude that the game is not strictly determined since there is no entry that is simultaneously the largest of the row minima and the smallest of the column maxima.

19. a.

$$\begin{array}{c c c c} & 1 & 2 & 3 \\ 1 & \begin{bmatrix} 2 & -3 & 4 \\ 2 & -3 & 4 & -5 \\ 3 & 4 & -5 & 6 \end{bmatrix} \end{array}$$

b. From

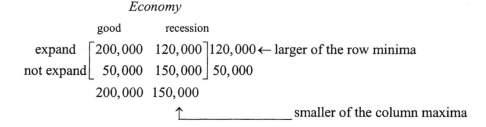

we conclude that the maximin strategy for Robin is to play row 1 (extend 1 finger), whereas the minimax strategy for Cathy is to play column 1 or column 2 (extend 1 or 2 fingers).

21. a. The following is the payoff matrix for the game.

Economy

good   recession

$$\begin{array}{c} \text{expand} \\ \text{not expand} \end{array} \begin{bmatrix} 200,000 & 120,000 \\ 50,000 & 150,000 \end{bmatrix} \begin{array}{l} 120,000 \leftarrow \text{larger of the row minima} \\ 50,000 \end{array}$$

200,000   150,000

↑_____ smaller of the column maxima

b. The row player's (management) minimax strategy is to play row 1, that is, to expand its line of conventional speakers.

23. a. The following is the payoff matrix for this game.

$$
\begin{array}{c}
\textit{Charley's} \\
\begin{array}{ccc} R & H & L \end{array}
\end{array}
$$

$$
\textit{Roland's} \begin{array}{c} R \\ H \\ L \end{array}
\left[\begin{array}{ccc}
3 & -1 & -3 \\
2 & 0 & -2 \\
5 & 2 & \boxed{1}
\end{array}\right]
\begin{array}{c} -3 \\ -2 \\ 1 \leftarrow \text{largest of the row minima} \end{array}
$$

$$
\begin{array}{ccc} 5 & 2 & 1 \end{array}
$$

$\uparrow\!\!\underline{\hspace{5cm}}$ smallest of the column maxima

b. From the payoff matrix, we see that the game is strictly determined.

c. If neither party is willing to lower their price, the payoff matrix would be

$$
\begin{array}{c}
\textit{Charley's} \\
\begin{array}{cc} R & C \end{array}
\end{array}
$$

$$
\textit{Roland's} \begin{array}{c} R \\ C \end{array}
\left[\begin{array}{cc}
3 & -1 \\
2 & \boxed{0}
\end{array}\right]
\begin{array}{c} -1 \\ 0 \leftarrow \text{larger of the row minima} \end{array}
$$

$$
\begin{array}{cc} 3 & 0 \end{array}
$$

$\uparrow\!\!\underline{\hspace{5cm}}$ smaller of the column maxima

and we see that the game is strictly determined, so that the optimal strategy for each barber is to charge his current price for a haircut.

25. True. This follows from the defintion.

EXERCISES 9.5, page 558

1. We compute

$$
E = PAQ = \begin{bmatrix} \frac{1}{2} & \frac{1}{2} \end{bmatrix} \begin{bmatrix} 3 & 1 \\ -4 & 2 \end{bmatrix} \begin{bmatrix} \frac{3}{5} \\ \frac{2}{5} \end{bmatrix}
$$

$$
= \begin{bmatrix} -\frac{1}{2} & \frac{3}{2} \end{bmatrix} \begin{bmatrix} \frac{3}{5} \\ \frac{2}{5} \end{bmatrix} = -\frac{3}{10} + \frac{6}{10} = \frac{3}{10}.
$$

Thus, in the long run, the row player may be expected to win 0.3 units in each play of the game.

3. We compute

$$
E = PAQ = \begin{bmatrix} \frac{1}{3} & \frac{2}{3} \end{bmatrix} \begin{bmatrix} -4 & 3 \\ 2 & 1 \end{bmatrix} \begin{bmatrix} \frac{3}{4} \\ \frac{1}{4} \end{bmatrix} = \begin{bmatrix} 0 & \frac{5}{3} \end{bmatrix} \begin{bmatrix} \frac{3}{4} \\ \frac{1}{4} \end{bmatrix} = \frac{5}{12}.
$$

Thus, in the long run, the row player may be expected to win 0.4167 units in each play of the game.

5. We compute

$$E = PAQ = \begin{bmatrix} .02 & .06 & .02 \end{bmatrix} \begin{bmatrix} 2 & 0 & -2 \\ 1 & -1 & 3 \\ 2 & 1 & -4 \end{bmatrix} \begin{bmatrix} 0.2 \\ 0.6 \\ 0.2 \end{bmatrix} = \begin{bmatrix} 1.4 & -0.4 & 0.6 \end{bmatrix} \begin{bmatrix} 0.2 \\ 0.6 \\ 0.2 \end{bmatrix} = 0.16.$$

7. a. $E = PAQ = \begin{bmatrix} 1 & 0 \end{bmatrix} \begin{bmatrix} 1 & -2 \\ -2 & 3 \end{bmatrix} \begin{bmatrix} 1 \\ 0 \end{bmatrix} = \begin{bmatrix} 1 & -2 \end{bmatrix} \begin{bmatrix} 1 \\ 0 \end{bmatrix} = 1$

   b. $E = PAQ = \begin{bmatrix} 0 & 1 \end{bmatrix} \begin{bmatrix} 1 & -2 \\ -2 & 3 \end{bmatrix} \begin{bmatrix} 1 \\ 0 \end{bmatrix} = \begin{bmatrix} -2 & 3 \end{bmatrix} \begin{bmatrix} 1 \\ 0 \end{bmatrix} = -2.$

   c.

   $$E = PAQ = \begin{bmatrix} \frac{1}{2} & \frac{1}{2} \end{bmatrix} \begin{bmatrix} 1 & -2 \\ -2 & 3 \end{bmatrix} \begin{bmatrix} \frac{1}{2} \\ \frac{1}{2} \end{bmatrix} = \begin{bmatrix} -\frac{1}{2} & \frac{1}{2} \end{bmatrix} \begin{bmatrix} \frac{1}{2} \\ \frac{1}{2} \end{bmatrix} = 0.$$

   d.

   $$E = PAQ = \begin{bmatrix} 0.5 & 0.5 \end{bmatrix} \begin{bmatrix} 1 & -2 \\ -2 & 3 \end{bmatrix} \begin{bmatrix} 0.8 \\ 0.2 \end{bmatrix} = \begin{bmatrix} -0.5 & 0.5 \end{bmatrix} \begin{bmatrix} 0.8 \\ 0.2 \end{bmatrix} = -0.3.$$

   (a) is the most advantageous.

9. a. From the payoff matrix

   $$\begin{bmatrix} -3 & 3 & 2 \\ -3 & 1 & 1 \\ 1 & -2 & 1 \end{bmatrix} \begin{matrix} -3 \\ -3 \\ -2 \leftarrow \end{matrix} \quad \text{largest of the row minima}$$

   $$\begin{matrix} 1 & 3 & 2 \\ \uparrow \end{matrix}$$

   smallest of the column maxima

   we see that the expected payoff to a row player using the minimax strategy is 1.

   b. The expected payoff is given by

   $$E = PAQ = \begin{bmatrix} .25 & .25 & .5 \end{bmatrix} \begin{bmatrix} -3 & 3 & 2 \\ -3 & 1 & 1 \\ 1 & -2 & 1 \end{bmatrix} \begin{bmatrix} .6 \\ .2 \\ .2 \end{bmatrix} = -0.35$$

c. The minimax strategy (part (a)) is the better strategy for the row player.

11. The game under consideration has no saddle point and is accordingly nonstrictly determined. Using the formulas for determing the optimal mixed strategies for a 2×2 game with $a = 4$, $b = 1$, $c = 2$, and $d = 3$, we find that

$$p_1 = \frac{d-c}{a+d-b-c} = \frac{3-2}{4+3-1-2} = \frac{1}{4}$$

$$p_2 = 1 - p_1 = 1 - \frac{1}{4} = \frac{3}{4},$$

so that the row player's optimal mixed strategy is given by

$$P = \begin{bmatrix} \frac{1}{4} & \frac{3}{4} \end{bmatrix}.$$

Next, we compute

$$q_1 = \frac{d-b}{a+d-b-c} = \frac{3-1}{4+3-1-2} = \frac{2}{4} = \frac{1}{2}.$$

$$q_2 = 1 - q_1 = 1 - \frac{1}{2} = \frac{1}{2}.$$

Thus, the optimal strategy for the column player is given by

$$Q = \begin{bmatrix} \frac{1}{2} \\ \frac{1}{2} \end{bmatrix}.$$

To determine whether the game favors one player over the other, we compute the expected value of the game which is given by

$$E = \frac{ad-bc}{a+d-b-c} = \frac{(4)(3)-(1)(2)}{4+3-1-2} = \frac{10}{4} = \frac{5}{2},$$

or 5/2 units for each play of the game. These results imply that the game favors the row player.

13. Since the game is not strictly determined, we use the formulas for determining the optimal mixed strategies for a 2×2 game. We obtain

$$p_1 = \frac{d-c}{a+d-b-c} = \frac{-3-1}{-1-3-2-1} = \frac{4}{7}; \quad p_2 = 1 - p_1 = 1 - \frac{4}{7} = \frac{3}{7}.$$

Thus, the optimal mixed strategy for the row player is given by

$$P = \begin{bmatrix} \frac{4}{7} & \frac{3}{7} \end{bmatrix}.$$

To find the optimal mixed strategy for the column player, we compute

$$q_1 = \frac{d-b}{a+d-b-c} = \frac{-3-2}{-1-3-2-1} = \frac{5}{7}; \quad q_2 = 1 - q_1 = 1 - \frac{5}{7} = \frac{2}{7}.$$

Hence, $Q = \begin{bmatrix} \frac{5}{7} \\ \frac{2}{7} \end{bmatrix}$. The expected value of the game is given by

$$E = \frac{ad - bc}{a + d - b - c} = \frac{(-1)(-3) - (2)(1)}{-1 - 3 - 2 - 1} = -\frac{1}{7}.$$

Since the value of the game is negative, we conclude that the game favors the column player.

15. Since the game is not strictly determined, we use the formulas for determining the optimal mixed strategies for a 2×2 game. We obtain

$$P = \begin{bmatrix} \frac{1}{2} & \frac{1}{2} \end{bmatrix} \quad \text{and} \quad Q = \begin{bmatrix} \frac{1}{4} \\ \frac{3}{4} \end{bmatrix}$$

and $E = PAQ = -5$. We conclude that the game favors the column player.

17. a. Since the game is not strictly determined, we employ the formulas for determining the optimal mixed strategies for a 2×2 game. We find that

$$p_1 = \frac{d - c}{a + d - b - c} = \frac{1 - (-2)}{4 + 1 - (-2) - (-2)} = \frac{3}{9} = \frac{1}{3}$$

$$p_2 = 1 - p_1 = 1 - \frac{1}{3} = \frac{2}{3},$$

so that Richie's optimal mixed strategy is given by

$$P = \begin{bmatrix} \frac{1}{3} & \frac{2}{3} \end{bmatrix}.$$

Next, we compute

$$q_1 = \frac{d - b}{a + d - b - c} = \frac{1 - (-2)}{4 + 1 - (-2) - (-2)} = \frac{3}{9} = \frac{1}{3}; \quad q_2 = 1 - q_1 = 1 - \frac{1}{3} = \frac{2}{3}.$$

Thus, Chuck's optimal strategy is given by $Q = \begin{bmatrix} \frac{1}{3} \\ \frac{2}{3} \end{bmatrix}$.

b. The expected value of the game is given by

$$E = \frac{ad - bc}{a + d - b - c} = \frac{(4)(1) - (-2)(-2)}{4 + 1 - (-2) - (-2)} = 0$$

and conclude that in the long run the game will end in a draw.

19. a. The required payoff matrix for this game is given by

Expanding economy   Economic recession

$$\begin{array}{c} \textit{Hotel stock} \\ \textit{Brewery stock} \end{array} \begin{bmatrix} 25 & -5 \\ 10 & 15 \end{bmatrix}$$

Since the game is not strictly determined, we use the formulas for finding the optimal mixed strategies for a 2×2 nonstrictly determined game. Then

$$p_1 = \frac{d-c}{a+d-b-c} = \frac{15-10}{25+15+5-10} = \frac{5}{35} = \frac{1}{7}; \; p_2 = 1 - p_2 = 1 - \frac{1}{7} = \frac{6}{7},$$

so that the Maxwell's optimal mixed strategy is $P = \begin{bmatrix} \frac{1}{7} & \frac{6}{7} \end{bmatrix}$

Thus, the Maxwells should invest (1/7)($40,000) = $5714 in hotel stocks and (6/7)($40,000) = $34,286 in brewery stocks.

b. The profit that the Maxwell's can expect to make is given by

$$E = \frac{ad-bc}{a+d-b-c} = \frac{(25)(15)-(-5)(10)}{35} = \frac{425}{35} = 12.1429.$$

We conclude that the Maxwell's will realize a profit of

$$(0.12143)(\$40,000) \approx \$4857$$

by employing their optimal mixed strategy.

21. a. The required payoff matrix for this game is given by

C

$$R \begin{array}{c} \\ N \\ F \\ \\ \end{array} \begin{array}{cc} N & F \\ \begin{bmatrix} .48 & .65 \\ .50 & .45 \end{bmatrix} & \begin{array}{c} .48 \\ .45 \end{array} \\ .50 \;\; .65 & \end{array} \qquad \begin{array}{l} N = \text{local newspaper} \\ F = \text{flyer} \end{array}$$

Since there is no saddle point, we conclude that the game is not strictly determined.
b. Employing the formulas for finding the optimal mixed strategies for a 2×2 nonstrictly determined game, we find that

$$p_1 = \frac{d-c}{a+d-b-c} = \frac{.45-.50}{.48+.45-.65-.50} = .227$$

$$p_2 = 1 - p_1 = 1 - .227 = .773.$$

$$q_1 = \frac{d-b}{a+d-b-c} = \frac{.45-.65}{.48+.45-.65-.50} = .909; \; q_2 = 1 - q_1 = 1 - .909 = .091.$$

We conclude that Dr. Russell's strategy is given by $P = [.227 \quad .773]$

and Dr. Carlton's strategy is given by $Q = \begin{bmatrix} .909 \\ .091 \end{bmatrix}$.

Also, Dr. Russell should place 22.7% of his advertisements in the local newspaper

and 77.3% in fliers; whereas, Dr. Carlton should place 90.9% of his advertisements in the local newspaper and 9.1% of his advertisements in fliers.

23. The optimal strategies for the row and column players are

$$P = [p_1 \quad p_2]$$

where

$$p_1 = \frac{d-c}{a+d-b-c} \quad \text{and}$$

$$p_2 = 1 - p_1 = 1 - \frac{d-c}{a+d-b-c} = \frac{a-b}{a+d-b-c}$$

and $\quad Q = \begin{bmatrix} q_1 \\ q_2 \end{bmatrix} \quad$ where $\quad q_1 = \frac{d-b}{a+d-b-c}$

and $\quad q_2 = 1 - q_1 = 1 - \frac{d-b}{a+d-b-c} = \frac{a-c}{a+d-b-c.}$

Therefore, the expected value of the game is

$$E = PAQ = [p_1 \quad p_2] \begin{bmatrix} a & b \\ c & d \end{bmatrix} \begin{bmatrix} q_1 \\ q_2 \end{bmatrix} = [ap_1 + cp_2 \quad bp_1 + dp_2] \begin{bmatrix} q_1 \\ q_2 \end{bmatrix}$$

$$= (ap_1 + cp_2)q_1 + (bp_1 + dp_2)q_2$$

$$= \left[\left[\frac{a(d-c)}{a+d-b-c} \right] + \left[\frac{c(a-b)}{a+d-b-c} \right] \right] \left[\frac{d-b}{a+d-b-c} \right]$$

$$+ \left[\left[\frac{b(d-c)}{a+d-b-c} \right] + \left[\frac{d(a-b)}{a+d-b-c} \right] \right] \left[\frac{(a-c)}{a+d-b-c} \right]$$

$$= \frac{(ad-bc)(d-b)}{(a+d-b-c)^2} + \frac{(ad-bc)(a-c)}{(a+d-b-c)^2}$$

$$= \frac{(ad-bc)(a+d-b-c)}{(a+d-b-c)^2} = \frac{ad-bc}{a+d-b-c},$$

as was to be shown.

1. Since the entries $a_{12} = -2$ and $a_{22} = -8$ are negative, the given matrix is not stochastic and is hence not a regular stochastic matrix.

3. $T^2 = \begin{bmatrix} \frac{1}{2} & 0 & \frac{1}{3} \\ 0 & 0 & \frac{1}{3} \\ \frac{1}{2} & 1 & \frac{1}{3} \end{bmatrix} \begin{bmatrix} \frac{1}{2} & 0 & \frac{1}{3} \\ 0 & 0 & \frac{1}{3} \\ \frac{1}{2} & 1 & \frac{1}{3} \end{bmatrix} = \begin{bmatrix} \frac{5}{12} & \frac{1}{3} & \frac{5}{18} \\ \frac{1}{6} & \frac{1}{3} & \frac{1}{9} \\ \frac{5}{12} & \frac{1}{3} & \frac{11}{18} \end{bmatrix}$ and so the matrix is regular.

5. $X_1 = \begin{bmatrix} 0 & \frac{1}{4} & \frac{3}{5} \\ \frac{2}{5} & \frac{1}{2} & \frac{1}{5} \\ \frac{3}{5} & \frac{1}{4} & \frac{1}{5} \end{bmatrix} \begin{bmatrix} \frac{1}{2} \\ \frac{1}{2} \\ 0 \end{bmatrix} = \begin{bmatrix} \frac{1}{8} \\ \frac{9}{20} \\ \frac{17}{40} \end{bmatrix}$. $X_2 = \begin{bmatrix} 0 & \frac{1}{4} & \frac{3}{5} \\ \frac{2}{5} & \frac{1}{2} & \frac{1}{5} \\ \frac{3}{5} & \frac{1}{4} & \frac{1}{5} \end{bmatrix} \begin{bmatrix} \frac{1}{8} \\ \frac{9}{20} \\ \frac{17}{40} \end{bmatrix} = \begin{bmatrix} \frac{147}{400} \\ \frac{9}{25} \\ \frac{109}{400} \end{bmatrix} = \begin{bmatrix} .3675 \\ .36 \\ .2725 \end{bmatrix}$.

7. This is an absorbing matrix since state 1 is an absorbing state and it is possible to go from any nonabsorbing state to state 1.

9. This is not an absorbing stochastic matrix since there is no absorbing matrix.

11. We solve the matrix equation
$$\begin{bmatrix} .6 & .3 \\ .4 & .7 \end{bmatrix} \begin{bmatrix} x \\ y \end{bmatrix} = \begin{bmatrix} x \\ y \end{bmatrix}$$
or equivalently, the system of equations
$$-.4x + .3y = 0$$
$$.4x - .3y = 0$$
$$x + y = 1$$
Solving this system of equations, we find the steady-state distribution vector to be $\begin{bmatrix} \frac{3}{7} \\ \frac{4}{7} \end{bmatrix}$ and the steady-state matrix to be $\begin{bmatrix} \frac{3}{7} & \frac{3}{7} \\ \frac{4}{7} & \frac{4}{7} \end{bmatrix}$.

13. We solve the system
$$\begin{bmatrix} .6 & .4 & .3 \\ .2 & .2 & .2 \\ .2 & .4 & .5 \end{bmatrix} \begin{bmatrix} x \\ y \\ z \end{bmatrix} = \begin{bmatrix} x \\ y \\ z \end{bmatrix}$$

together with the equation $x + y + z = 1$, or equivalently, the system

$$.6x + .4y + .3z = x$$
$$.2x + .2y + .2z = y$$
$$.2x + .4y + .5z = z$$
$$x + y + z = 1$$

upon solving the system, we find the $x = .457$, $y = .20$, and $z = .343$,

and the steady-state distribution vector is given by $\begin{bmatrix} .457 \\ .20 \\ .343 \end{bmatrix}$ and the steady-state matrix

is $\begin{bmatrix} .457 & .457 & .457 \\ .20 & .20 & .20 \\ .343 & .343 & .343 \end{bmatrix}$.

15. a. The transition matrix for the Markov Chain is given by

$$T = \begin{array}{c} \\ A \\ U \\ N \end{array} \begin{array}{ccc} A & U & N \\ \begin{bmatrix} .85 & 0 & .10 \\ .10 & .95 & .05 \\ .05 & .05 & .85 \end{bmatrix} \end{array}.$$

The probability vector describing the distribution of land 10 years ago is given by

$$\begin{array}{c} A \\ U \\ N \end{array} \begin{bmatrix} .50 \\ .15 \\ .35 \end{bmatrix}.$$

To find the required probability vector, we compute

$$TX_0 = \begin{bmatrix} .85 & 0 & .10 \\ .10 & .95 & .05 \\ .05 & .05 & .85 \end{bmatrix} \begin{bmatrix} .50 \\ .15 \\ .35 \end{bmatrix} = \begin{bmatrix} .46 \\ .21 \\ .33 \end{bmatrix}$$

$$TX_1 = \begin{bmatrix} .85 & 0 & .10 \\ .10 & .95 & .05 \\ .05 & .05 & .85 \end{bmatrix} \begin{bmatrix} .46 \\ .21 \\ .33 \end{bmatrix} = \begin{bmatrix} .424 \\ .262 \\ .314 \end{bmatrix}.$$

Thus, the probability vector describing the distribution of land 10 years from now is

$$\begin{bmatrix} .424 \\ .262 \\ .314 \end{bmatrix}.$$

17. From

$$\begin{bmatrix} 1 & 2 \\ 3 & 5 \\ \boxed{4} & 6 \end{bmatrix} \begin{matrix} 1 \\ 3 \\ 4 \end{matrix} \leftarrow \text{largest of the row minima}$$

       4 6

     ↑_____smaller of the column maxima

we see that the game is strictly determined, and
a. the saddle point is 4.
b. the optimum strategy for the row player is to play row 3, whereas the optimum strategy for the column player is to play column 1.
c. the value of the game is 4.
d. the game favors the row player.

19. We first determine the largest of the row minima nd the smallest of the column maxima and display these elements as follows:

$$\begin{bmatrix} \boxed{1} & 3 & 6 \\ -2 & 4 & 3 \\ -5 & -4 & -2 \end{bmatrix} \begin{matrix} 1 \\ -2 \\ -5 \end{matrix} \leftarrow \text{the largest of the row minima}$$

      1   4   6

     ↑_____ the smallest of the column maxima

The entry $a_{11} = 1$ is the saddle point of the game and we conclude that the game is strictly determined. The row player's optimal strategy is to play row 1 and the column player's optimal strategy is to play column 1. The value of the game is 1 and the game favors the row player.

21. We compute $E = PAQ = \begin{bmatrix} \frac{1}{2} & \frac{1}{2} \end{bmatrix} \begin{bmatrix} 4 & 8 \\ 6 & -12 \end{bmatrix} \begin{bmatrix} \frac{1}{4} \\ \frac{3}{4} \end{bmatrix} = -\frac{1}{4}.$

23. We compute

$$E = PAQ = \begin{bmatrix} .2 & .4 & .4 \end{bmatrix} \begin{bmatrix} 3 & -1 & 2 \\ 1 & 2 & 4 \\ -2 & 3 & 6 \end{bmatrix} \begin{bmatrix} .2 \\ .6 \\ .2 \end{bmatrix} = \begin{bmatrix} .2 & 1.8 & 4.4 \end{bmatrix} \begin{bmatrix} .2 \\ .6 \\ .2 \end{bmatrix} = 2.$$

The expected payoff for the game is 2.

25. The game under consideration has no saddle point and is accordingly nonstrictly determined. Using the formulas for determing the optimal mixed strategies for a 2×2 game with $a = 1$, $b = -2$, $c = 0$, and $d = 3$, we find that

$$p_1 = \frac{d - c}{a + d - b - c} = \frac{3 - 0}{1 + 3 + 2 - 0} = \frac{3}{6} = \frac{1}{2}; \quad p_2 = 1 - p_1 = 1 - \frac{1}{2} = \frac{1}{2}$$

so that the row player's optimal mixed strategy is given by $P = \begin{bmatrix} \frac{1}{2} & \frac{1}{2} \end{bmatrix}$.

Next, we compute $q_1 = \frac{d - b}{a + d - b - c} = \frac{3 + 2}{1 + 3 + 2 - 0} = \frac{5}{6}; \quad q_2 = 1 - q_1 = 1 - \frac{5}{6} = \frac{1}{6}.$

Thus, the optimal strategy for the column player is given by $Q = \begin{bmatrix} \frac{5}{6} \\ \frac{1}{6} \end{bmatrix}$.

To determine whether the game favors one player over the other, we compute the expected value of the game which is given by

$$E = \frac{ad - bc}{a + d - b - c} = \frac{(1)(3) - (-2)(0)}{1 + 3 + 2 - 0} = \frac{3}{6} = \frac{1}{2}$$

or 1/2 units for each play of the game. These results imply that the game favors the row player.

27. Using the formulas for the optimal strategies in a 2×2 nonstrictly determined game,

we see that the optimal mixed strategy for the row player is $P = \begin{bmatrix} \frac{1}{10} & \frac{9}{10} \end{bmatrix}$

and the optimal mixed strategy for the column player is $Q = \begin{bmatrix} \frac{4}{5} \\ \frac{1}{5} \end{bmatrix}$. The value of the

game is $PAQ = [1.2]$ and the game favors the row player.

29. a. The required payoff matrix is given by

$$\begin{array}{cc} & \text{Record World} \\ & \begin{array}{cc} \$7 & \$8 \end{array} \\ \text{Disco-Mart} \begin{array}{c} \$7 \\ \$8 \end{array} & \begin{bmatrix} .5 & .7 \\ .4 & .5 \end{bmatrix} \begin{array}{c} .5 \\ .4 \end{array} \\ & \begin{array}{cc} .5 & .7 \end{array} \end{array}$$

b.   Upon finding the larger of the row minima and the smaller of the column maxima, we see that the entry $a_{11} = .5$ is a saddle point, and , consequently, that the game is strictly determined. Thus, the optimal price that each company should sell the compact disc label for is $7.